SpringerBriefs in Electrical and Computer Engineering

More information about this series at http://www.springer.com/series/10059

Aiqing Zhang • Liang Zhou • Lei Wang

# Security-Aware Device-to-Device Communications Underlaying Cellular Networks

Aiqing Zhang
College of Telecommunications
 and Information Engineering
Nanjing University of Posts
 and Telecommunications
Nanjing, Jiangsu, China

Lei Wang
Key Lab of Broadband Wireless
 Communication and Sensor Network
 Technology
Nanjing University of Posts
 and Telecommunications
Nanjing, Jiangsu, China

Liang Zhou
Key Lab of Broadband Wireless
 Communication and Sensor Network
 Technology
Nanjing University of Posts
 and Telecommunications
Nanjing, Jiangsu, China

ISSN 2191-8112           ISSN 2191-8120   (electronic)
SpringerBriefs in Electrical and Computer Engineering
ISBN 978-3-319-32457-9        ISBN 978-3-319-32458-6   (eBook)
DOI 10.1007/978-3-319-32458-6

Library of Congress Control Number: 2016939117

Printed on acid-free paper

This Springer imprint is published by Springer Nature
The registered company is Springer International Publishing AG Switzerland

# Preface

With an explosive growth of mobile population and wireless multimedia data, it is increasingly necessary and important to off-load the busy traffic of cellular networks. As device-to-device (D2D) communication is a promising data off-loading solution in long-term evolution (LTE) networks, it has received a substantial amount of interest in communication and network research communities recently. D2D communications enable devices to communicate directly; thus, they have the advantages of improving resource utilization, enhancing user's throughput, extending battery lifetime, etc. However, due to the open nature of D2D communications, they face two substantial technical challenges when it applies to large-scale applications, that is, security and availability. The objective of this book is to present systematical mechanisms to realize system security and availability for D2D communications.

Chapter 1 gives an overview of D2D communications, including their architecture, characteristics, application scenarios, and open topics. Then, the security issues are presented from the aspects of security architectures, threat model, and security requirements. Finally, the organization of the book is depicted for a comprehensive understanding of the book.

In Chap. 2, we propose a secure data sharing protocol, which merges the advantages of public key cryptography and symmetric encryption, to achieve data security in D2D communications. Specifically, digital signature combining with mutual authentication between the evolved NodeB (eNB) and users guarantees the entity authentication, data authority and integrity, and transmission non-repudiation as well as traceability. Meanwhile, data confidentiality is achieved through symmetric encryption. The key hint transmission between the eNB and user equipments (UEs) realizes reception non-repudiation. Free-riding attack is detected by keeping a record of the data sharing behaviors for the UEs in the eNB; thus, the system availability is improved.

In Chap. 3, an overview of physical-layer security is firstly given in terms of secrecy capacity, physical-layer key agreement, and physical-layer authentication. Then a joint framework involving both the physical- and application-layer security technologies is proposed for multimedia service over D2D communications. In this scheme, the scalable security service can be achieved without changing the current communication framework.

Furthermore, as the system availability largely depends on the cooperation degree of the users, Chap. 4 focuses on cooperation stimulation. As multimedia dominates the contents with quality of experience (QoE) as a key measurement, the cooperation stimulation mechanism is constructed for maximizing user QoE characterized by mean opinion score (MOS). In the proposed scheme, the users compute their transmitter MOS and receiver MOS and send them to the content provider (CP). Then, the CP formulates a weighted directed graph based on the network topology and connection MOS. By seeking 1-Factors for the graph, the content dissemination scheme is designed according to the 1-Factor with the maximum weight. Additionally, in order to realize cheat-proof, a debt mechanism is introduced in the scheme. Simulation results demonstrate that the proposed scheme gives due consideration to efficiency and fairness for content dissemination in D2D communications.

Finally, Chap. 5 summarizes this book and highlights the future research directions.

Nanjing, China                                                      Aiqing Zhang
October 2015                                                         Liang Zhou
                                                                       Lei Wang

# Acknowledgments

This work is partly supported by the National Natural Science Foundation of China (Grants No. 61322104, 61571240), Natural Science Foundation of Anhui Province (1608085QF138), University Natural Science Research Foundation of Anhui Province (KJ2015A105, KJ2015A092), and Key Projects of the Outstanding Young Talents Program in Universities of Anhui Province (gxyqZD2016027).

# Contents

# Chapter 1
# Introduction

**Abstract** With the steep growth of mobile population and wireless multimedia data in recent years, Device-to-Device (D2D) communications are promising data offloading solutions and spectrum efficiency enhancement methods. They have drawn considerable attention in communication research community. In this chapter, we firstly give an overview of D2D communications from the aspects of their characteristics, application scenarios, and research topics. Then, the security architectures, threat model, and security requirements of D2D communications are presented. Finally, the organization of the book is depicted for a comprehensive understanding for all the content.

## 1.1 Overview of D2D Communications

### 1.1.1 D2D Communications

Recent years have witnessed the explosive growth of mobile population and data traffic, which bring extremely demanding in terms of network resources and link capacity, as shown in Fig. 1.1. The current cellular network is pushed closely to its limit facing the overwhelming service request, thus both academia and industry put forward the next generation mobile communication system, referred to as 5G, in order to envision the variety requests for future networks [1, 3, 5].

Device-to-Device (D2D) communications, which allow two nearby devices to communicate directly with each other, is introduced in 5G to offload the overburden networks [2, 4]. Here the term device refers to a mobile phone, tablet, laptop, or any other portable wireless device with cellular connectivity. Generally, D2D communications fall into two classifications: *Inband* D2D communications and *Outband* D2D communications.

*Inband D2D Communications* Inband D2D communications work on licensed spectrum and they are further divided into underlay and overlay categories [6]. In underlay D2D communications, cellular and D2D communications share the

A. Zhang et al., *Security-Aware Device-to-Device Communications Underlaying Cellular Networks*, SpringerBriefs in Electrical and Computer Engineering, DOI 10.1007/978-3-319-32458-6_1

1

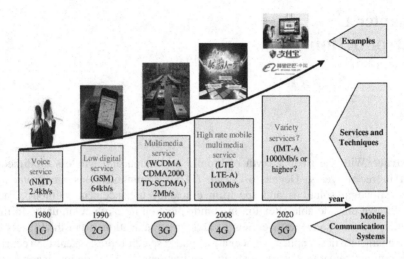

**Fig. 1.1** Evolution of mobile communications

same radio resources, thus they have the advantages of high spectral efficiency and easy QoS provisioning, while the interference management and resource allocation methods among D2D and cellular transmission are very challenging. In contrast, overlay D2D communications are given dedicated cellular resources which may waste cellular resources.

*Outband D2D Communications* Outband D2D communications, which exploit unlicensed spectrum to eliminate the interference between D2D and cellular links, can be further divided into controlled and autonomous communications. In autonomous outband D2D communications, cellular network leaves the D2D communications to the users just like the traditional wireless technologies, such as Wi-Fi Direct, ZigBee, or Bluetooth. The controlled outband D2D communications are under the control of cellular networks, which provide a planned deployment thus bringing in better user experiences [3].

In this book, we focus on D2D communications underlaying cellular networks. Figure 1.2 presents a system model of D2D communications.

### 1.1.2 Application Scenario

As D2D communications enable proximity devices to communicate directly, they have inherent characteristics, e.g., improving resource utilization, bringing hop gains, extending battery lifetime, enhancing system capacity, etc. [7, 8]. They are proposed to be widely used in *proximity service, machine-to-machine communications, and content dissemination.*

**Fig. 1.2** System model of D2D communications

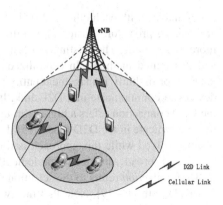

Specifically, D2D communications may provide proximity services ranging from *mobile multi-player gaming, mobile advertising, streaming services*, to *community services* [9]. Furthermore, D2D communications can be of critical use in natural disasters, e.g., an urgent communication network can be established using D2D functionality in a short time to replace the damaged communication networks and Internet infrastructures in an earthquake or hurricane [6]. D2D communications can also be used in machine-to-machine communications such as *Vehicle-to-Vehicle (V2V) communication* because of its low end-to-end delay. For example, it coordinates braking between vehicles for collision avoidance as well as providing information about the nearest car.

Another advantageous application of D2D communications is *data sharing or content dissemination* [10]. According to the prediction of Cisco [11], over two-thirds of the world's mobile data traffic will be video by 2018. Consequently, D2D or content dissemination is an important technology for offloading the busy traffic of cellular networks. Moreover, D2D multimedia content sharing provides good user quality of experience (QoE) as it has the advantages of low transmission delay and low energy consumption. Therefore, our book concentrates on the secure multimedia content transmission in D2D communications.

## *1.1.3 State-of-the-Art*

### 1.1.3.1 Physical-Layer Techniques

Currently, there are abundant researches on physical-layer techniques in D2D communications, including peer discovery, mode selection, resource allocation, power control, and interference management.

*Peer discovery* is the first procedure aiming to discover the presence of their peer D2D candidates and to identify whether candidate D2D pairs require to communicate with each other [12–15]. It is necessary to make such peer discovery and pairing procedures faster, more efficient in terms of energy consumption, and more user friendly. The existing works are classified into two categories: centralized and distributed methods. In centralized method, the packet data network (PDN) gateway or mobility management entity (MME) finds that it may be better for two devices to communicate in a D2D link. It then requests the eNB to check whether the D2D connection offers a higher throughput. If so, the eNB decides the two UEs to communicate in the D2D mode [15]. In the distributed approach, the base station is not involved while this approach is time and energy consuming due to the fact that the UE broadcasts identity periodically to inform others of its existence [14].

*Mode selection* refers to the selection between the traditional infrastructure path and D2D path [16–19]. Note that the switch between the two links is required to be seamless to the users for guaranteeing the user satisfaction. The key point of its designation lies in at what time scale mode selection should operate. As the radio conditions within the cell and between the D2D pairs may change rapidly, the time scale for mode selection cannot be too coarse. On the other hand, the control signaling required for mode selection should not be too frequent to avoid heavy overhead. Usually, mode selection is considered with resource allocation [18].

*Resource allocation* plays a key role in D2D communications [20–24]. It can be considered as an optimization mechanism aiming to improve the performances of the energy consumption, throughput, and interference by methods of scheduling, mode selection, and power control. Resource allocation needs to decide whether upper link (UL) or downlink (DL) physical resource blocks (PRBs) should be used for the D2D links, and which should be used in each direction for D2D communications. Generally, resources can be either distributively allocated by the UEs themselves or centrally determined by the eNB in LTE systems. The distributed approaches schedule a channel state aware maximal independent set at any given time slot based on the current traffic and channel condition. In the centralized resource allocation approaches, the eNB has full control over the resources allocated to each D2D link and needs to inform the D2D UEs of the scheduled resources for data transmission via physical downlink control channel (PDCCH).

*Power control* and *interference management* are usually realized by jointly considering mode selection and resource allocation [17, 25–32]. D2D links may find short time intervals and frequency proportions, where communication is feasible without causing harmful interference to the cellular network. The eNB may assign dedicated PRBs for D2D communications, where these resources are dynamically adjusted based on the temporal needs. But it could lead to inefficient use of the available resources. As D2D links reuse the same PRBs, which is allocated for the cellular links, the spectrum efficiency is increased. Additionally, various interference avoiding multiple-input-multiple output (MIMO) techniques can be combined with the advanced (network) coding schemes.

Technically, physical-layer approaches guarantee the D2D pairs meet in space, time, and frequency, and the links are connected timely. However, they don't concern the specific applications.

### 1.1.3.2   Content Dissemination

As content dissemination is an important application of D2D communications, it has drawn lots of attention recently [33–38]. Golrezaei et al. [33] shows that non-vanishing throughput per node can be achieved by distributed caching if there is a sufficient content reuse. Li and Wang [34] study analytic bounds of dissemination distance and hitting time in dynamic, intermittently connected networks. The authors in [35] find that the coded multi-cast gain and the spatial reuse gain do not cumulate in terms of the throughput scaling laws, somewhat counter-intuitively yet heuristic. Borst et al. [36] develops lightweight cooperative cache management algorithms in order to maximize the traffic volume served from caches and minimize the bandwidth cost. Antonopoulos et al. [37] proposes two energy-aware game-theoretic MAC strategies, providing guidelines for novel MAC protocol design in D2D communications. A mobile social network (MSN)-aided content dissemination is proposed in [38] by allowing the users to create a self-organized ad hoc network, aiming at studying dynamic content dissemination over an MSN.

Additionally, there are many works addressing content dissemination issues in other wireless networks [39–42, 44, 45]. In [40], the authors research media provision in the heterogeneous peer-to-peer (P2P)-based vehicular networks. They develop fully dynamic service schemes in order to maximize the total user satisfaction and achieve fairness to some extent. By proposing a media-aware satisfaction-fairness strategy, it ensures max–min satisfaction fairness among vehicles. The schemes are designed in a distributed manner, thus it is helpful to shed insights on the protocol design for an efficient data dissemination scheme in D2D communication systems. In [41], the authors explore how the service provider can allocate its bandwidth optimally to make the content as "fresh" as possible for the users, and specify the condition under which the system is highly scalable. This research provides a good direction for dynamic content sharing in D2D communications.

The authors in [42] employ a word-of-mouth demand evolution model [43] to represent the evolution of content interest to evaluate the benefits of a hybrid system that combines peer-to-peer and a centralized client–server approach. It models the file request as a function of time using the Bass diffusion model, denoting the temporal evolution of popular files. The results provide a guidance to understand how server provisioning affects performance of content distribution in D2D communication systems. Moreover, the contribution is heuristic for content dissemination mechanism design of D2D communications as the eNB plays the role of control center.

In [44] and [45], the content dissemination problem in vehicle ad hoc networks is addressed. Particularly, [45] studies a push-based content dissemination while [44] focuses on a pull-based content dissemination. Moreover, the authors investigate the dissemination capacity (DC) of the proposed scheme, which provides guidelines on choosing the system parameters thus maximizing the DC under different delivery ratio requirements.

These researches are pioneering works and shed lights on the content dissemination in D2D communications. However, they don't show how to guarantee security for D2D communications.

### 1.1.3.3   Security

As the connections happen directly between the proximity devices, D2D communications may subject to many security threats such as modification and fabrication of the data and violation of the user privacy. However, data security is the prerequisite in most of the D2D communication application scenarios such as V2V communication or content transmission. Thereafter, security is an important issue in D2D communications. Generally, the existing researches focus on secure content dissemination with energy considerations, as shown in Table 1.1.

Specifically, [46] provides an overview of the security architecture, threads, and requirements in D2D communications. Several potential solutions are proposed by reusing the existing security mechanisms. This research gives a guideline for realizing secure data sharing in D2D communications while not giving specific key agreement methods and authentication mechanisms. Panaousis et al. [47] investigates the best D2D network path to deliver a potentially malicious message from a source to a destination by solving a secure message delivery game. The scheme considers the energy costs and quality of service as well as security. Nobach and Hausheer [48] sketches a new approach towards decentralized energy- and privacy-aware D2D content delivery. In order to resist main-in-the-middle (MITM)

**Table 1.1**  State-of-the-art of secure D2D communications

| Items | Literatures | Security threats | Security methods | Security objectives |
|---|---|---|---|---|
| Secure content dissemination | [46] | Eavesdropping, impersonation | Cryptograph, authentication | Data confidentiality, reliability |
| | Panaousis et al. [47] | Fabrication | Transmission protocol | Malicious information detection |
| | Nobach and Hausheer [48] | Privacy violation | Transmission protocol | Privacy preservation |
| | Kwon et al. [49] | Main-in-the-middle attack, replay attacks | Cryptograph, authentication | Data integrity |
| Physical-layer security | [50] | Eavesdropping | Power control | Maximize secrecy capacity |
| | Zhang et al. [51] | Eavesdropping | Resource allocation | Maximize secrecy capacity |
| | Zhu et al. [52] | Eavesdropping | Power control | Secrecy outage probability |

and replay attacks in mobile multi-hop D2D communications, [49] proposes D2D authentication protocols with a secure initial key establishment using ciphertext policy attribute-based encryption (CP-ABE). In the proposed scheme, the communicating parties authenticate each other mutually and the link key is derived in a secure way for multi-hop D2D communications.

There are also some researches on physical-layer security [50–52] in D2D communications. Yue et al. derive the optimal transmission power to maximize the secrecy capacity of the cellular users in [50]. They also utilize the secrecy outage probability to depict the imperfect channel state information at the eavesdropper. This is a pioneer work on physical-layer security in D2D communications while the security capacity of D2D users is not investigated. Then D2D secrecy capacity is explored in [51], where the authors consider secrecy capacity for both cellular users and D2D users. Zhu et al. [52] derive the secrecy outage probability for the D2D and cellular users. As far as we know, this is the first work to derive the secrecy outage probability for the D2D communications in the presence of multi-antenna eavesdroppers.

Note that in the above studies, the application-layer security and physical-layer security are considered, respectively. Actually, scalable security services can be maximized within the given multimedia delivery deadlines by exploiting the security capacity and signal processing technologies at the physical layer and the authentication strategies at the application layer, which will be analyzed in this book.

### 1.1.3.4  Cooperation Stimulation

In D2D communications, the users work both as the servers and clients, thus the system availability largely depends on the cooperation degree of the users. However, selfish users are not voluntarily interested in transmitting contents to others without sufficient incentive because of the energy and bandwidth consumption. Consequently, cooperation stimulation turns to be another important research topics in D2D communications. Albeit there are abundant researches on cooperation stimulation in wireless networks such as mobile ad hoc network and MSN [53–59], very limited work has been proposed to address this issue in D2D communications up to this end [60, 61]. The authors develop a coalitional game-theoretic framework to devise social-tie-based cooperation strategies for D2D communications in [60] and [61], which sheds light on this research line in D2D communications.

The current cooperation stimulation strategies fall into the following three categories: reputation-based (RB), credit-based (CB), and resource-exchange-based (EB). In RB mechanisms, nodes monitor each other's behavior, evaluate each node's trustworthiness, cooperate with those which maintain good reputation, and detect misbehavior [53, 54]. CB systems treat packet forwarding or data sharing services as transactions so the nodes pay credit (or virtual currency) to the service providers [55, 56]. In EB mechanisms, two cooperative nodes act as a server as well as a potential client [57–59].

Considering the particular features of D2D communications, resource-exchange-based incentive mechanism is a superior choice. This is because it neither relies on the tamper-proof hardware nor monitoring other's behavior, which brings costs or privacy issues in D2D communications. Different from the previous resource-exchange-based mechanism [57, 58], in which QoE is not a concern, we propose a QoE-driven cooperation stimulation mechanism in this brief because QoE is an important concern in next generation mobile communication systems.

## 1.2  Security Issues

### 1.2.1  Security Architecture

The 3rd Generation Partnership Project (3GPP) has published Proximity Service (ProSe) high-level reference model [62] and security architecture [63]. Based on the proposition, Alam et al. propose the security architecture for D2D communications in LTE, as shown in Fig. 1.3. The proposed architecture is composed by five security feature groups against different attacks for different goals [3, 46]:

- Network access security (1): This group concerns the security interactions between users and access networks, and provides the protection for the radio access link.
- Network domain security (2): It enables secure data/control signal exchange among network elements and provides protection against attacks on wireline network.

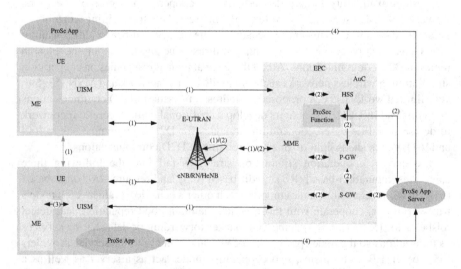

**Fig. 1.3** Security architecture for LTE proximity service [3, 46]

- User domain security (3): It provides protections for the access to the mobile station.
- Application domain security (4): It ensures the end-to-end security of the application on users and servers.
- Visibility and configuration security (5): This controls the availability and configuration of certain security service.

There are additional three functions for the ProSe, including ProSe function, ProSe apps server and apps, and Radio link between two UEs.

- ProSe function. This function is responsible for storing user-specific configurations, realizing peer discovery, and so on. It interacts with the mobile management element (MME), home subscriber server (HSS), ProSe application server, and other network element/ module related to ProSe via wireline.
- ProSe apps server and apps. In the IP network, the server has a logical link to the user apps on UE.
- Radio link between two UEs. This radio link is in the LTE-A frequency band, and managed by the network (if available). It is dedicated to D2D communications.

From the figure, it can be found that a direct radio link between the UEs is the most vulnerable. We mainly consider the security threats and requirements of the UEs in this brief.

## 1.2.2  Security Threats

Due to the open nature of D2D communications, they may introduce various security vulnerabilities.

*Eavesdropping* Eavesdropping is in essence a hidden danger in wireless communications. The unauthorized users may eavesdrop the in-transmit message as long as they work on the same frequency as the D2D communications. This threat is required to be taken against in most of the applications as it is content sensitive. Usually, cryptography approach is adopted to realize the data confidentiality.

*Fabrication* The adversary may modify the source data or forge the content and send the fake information to the other users. Alternatively, the contaminated devices may infect their neighbors without conscious, which brings chaos even damage to the system. The users may append digital signature with the data to guarantee data integrity.

*Free-Riding Attack* Due to the energy consumption of the data transmission process, some selfish UEs may not be willing to send contents to others while receiving its demanding data from their peers. This behavior is referred to as free-riding attack, which breaks fairness and reduces the system availability in D2D communications.

*Privacy Violation* The devices are controlled by the users in D2D communications, thus some privacy-sensitive information such as the user' s name and position can be intentionally derived by the attackers, bringing jeopardize to personal privacy.

*Denial-of-Service (DoS) Attack* The adversary sends large bulks of irrelevant messages to take up the channel and consume the computational resources of the other users. It is characterized by an explicit attempt by attackers to prevent legitimate users of a service from using that service.

## 1.2.3 Security Requirements

In order to mitigate most of the potential threats, a well-developed security scheme should meet the following security requirements [64]:

*Data Confidentiality* Confidentiality ensures that the data is well protected and not revealed to the unauthorized users. Even though the message may be eavesdropped by the unintended receiver, the confidentiality objective is expected to thwart the thief from getting any useful information.

*Data Authority and Correctness* The original authorized message should not be altered during the content dissemination and can be detected by the receiver if modification incident happens. Consequently, any false data fabricated by the adversary should be easily found out upon reception.

*Entity Authentication* Entity authentication techniques verify the identity of the devices in communications and distinguish legitimate users from unauthorized ones.

*No-Repudiation* No-repudiation includes transmission no-repudiation and receive no-repudiation, which prevent legitimate users from denying transmission or reception of message. To achieve this security objective, digital signature is usually adopted, which is efficient in transmission no-reputation while fails to deal with reception no-reputation. Usually, an additional scheme is required to be introduced to realize receive no-repudiation.

*Privacy Preservation* Privacy preservation is a critical security requirement for D2D communication system in practical implementation and commercialization. In particular, it is necessary to prevent users (both transmitters and receivers) from obtaining others' privacy information, such as identity, SIM card number, and position. However, the privacy protection in D2D communications should be conditional, i.e., senders and receivers are anonymous to each other while traceable by the trust authority (TA). With traceability, the TA is supposed to reveal the source ID of false messages.

*Availability* Users may be frustrated if services become temporarily unavailable due to the attacks, such as free-riding. Moreover, users might also be unhappy if they suffer from a long waiting time for sharing the information. Thus, it is necessary to develop some elaborate and carefully designed cooperation stimulation mechanism to resist free-riding in D2D communications [65, 66].

## 1.3 Organization

This book focuses on security and availability for D2D communications underlaying cellular networks. The remainder of this book is organized as follows. Chapter 2 presents a secure data sharing protocol which merges the advantages of public key cryptography and symmetric encryption to achieve data security in D2D communications. In Chap. 3, a joint physical–application layer security technology is proposed for multimedia service over D2D communications to efficiently utilize the available network resources. In order to realize system availability, a QoE-driven cooperation stimulation scheme is present in Chap. 4. Finally, we conclude the book and present the future research directions in Chap. 5.

## References

1. Condoluci M, Dohler M, Araniti G, Molinaro A, Zheng K (2015) Toward 5G DenseNets: architectural advances for effective machine-type communications over femtocells. IEEE Commun Mag 53(1):134–141
2. Wu D, Wang J, Hu R, Cai Y, Zhou L (2014) The role of mobility for D2D communications in LTE-advanced networks: energy vs. bandwidth efficiency. IEEE Wirel Commun 21(4):66–71
3. Mumtaz S, Mohammed K, Huq S, Rodriguez J (2014) Direct mobile-to-mobile communications: paradigm for 5G. IEEE Wirel Commun (21) 10:14–23
4. Zhu, D, Swindlehurst, A, Fakoorian, S et al (2014) Device-to-device communications: the physical-layer security advantage. In: IEEE international conference on acoustic, speech and signal processing, Florence, Italy, May 4–9, pp 1606–1610
5. Tehrani MN, Uysal M, Yanikomeroglu H (2014) Device-to-device communication in 5G cellular networks: challenges, solutions, and future directions. IEEE Commun Mag (52)5:86–92
6. Asadi A, Wang Q, Mancuso V (2014) A survey on device-to-device communication in cellular networks. IEEE Commun Surv Tutorials 16(4):1801–1819
7. Zhou L, Wu D, Zheng B et al (2014) Joint physical-application layer security for wireless multimedia delivery, IEEE Commun Mag 52(3):66–72
8. Ryu S, Park S, Park N, Chung S (2013) Development of device-to-device communication based new mobile proximity multimedia service business models. In: IEEE international conference on multimedia and expo workshops.
9. Mumtaz S, Rodriguez J (2014) Smart device to smart device communication. Springer, Cham
10. Zhang A, Chen J, Zhou L, Yu S (2015) Graph theory based QoE-driven cooperation stimulation for content dissemination in device-to-device communication. IEEE Trans Emerg Top Comput. doi: 10.1109/TETC.2015.2430816
11. Cisco visual network index: global mobile data traffic forecast update, 2013–2018
12. Zhang B, Li Y, Jin D, Hui P, Han Z (2015) Social-aware peer discovery for D2D communications underlaying cellular networks. IEEE Trans Wirel Commun 14(5):2426–2439
13. Prasad A, Kunz A, Velev G, Samdanis K, Song J (2014) Energy efficient D2D discovery for proximity services in 3GPP. IEEE Veh Technol Mag 9(4): 40–50
14. Chao S, Lee H, Chou C, Wei H (2013) Bio-inspired proximity discovery and synchronization for D2D communications. IEEE Commun Lett 17(12):2300–2303
15. Fodor G, Dahlman E, Mildh G et al (2012) Design aspects of network assisted device-to-device communications. IEEE Commun Mag 50(3):170–177
16. Zhu K, Hossain E (2015) Joint mode selection and spectrum partitioning for device-to-device communication: a dynamic Stackelberg game. IEEE Trans Wirel Commun 14(3):1406–1420

17. Elsawy H, Hossain E, Alouini, MS (2014) Analytical modeling of mode selection and power control for underlay D2D communication in cellular networks. IEEE Trans Commun 62(11):4147–4161

18. Wen S, Zhu X, Zhang X, Yang D (2013) QoS-aware mode selection and resource allocation scheme for device-to-device (D2D) communication in cellular networks. In: IEEE internet conference on communications

19. Min H, Seo W, Lee J, Park S, Hong D (2011) Reliability improvement using receive mode selection in the Device-to-Device uplink period underlaying cellular networks. IEEE Trans Wirel Commun 10(2):413–418

20. Wu D, Cai Y, Hu R Q, Qian Y (2015) Dynamic distributed resource sharing for mobile D2D communications. IEEE Trans Wirel Commun. doi: 10.1109/TWC.2015.2438292

21. Wei L, Hu R Q, Qian Y, Wu G (2015) Energy-efficiency and spectrum-efficiency of multi-hop device-to-device communications underlaying cellular networks. IEEE Trans Veh Technol. doi: 10.1109/TVT.2015.2389823

22. Wu D, Wang J, Hu R Q, Cai Y, Zhou L (2014) Energy-efficient resource sharing for mobile device-to-device multimedia communications. IEEE Trans Veh Technol 63(5):2093–2103

23. Wu D, Zhou L, Cai Y (2012) Energy-efficient resource allocation for uplink orthogonal frequency division multiple access systems using correlated equilibrium. IET Commun 6(6):659–667

24. Malandrino F, Limani Z, Casetti C, Chiasserini C (2015) Interference-aware downlink and uplink resource allocation in HetNets with D2D support. IEEE Trans Wirel Commun 14(5):2729–2741

25. Lee N, Lin X, Andrews J, Heath R (2015) Power control for D2D underlaid cellular networks: modeling, algorithms, and analysis. IEEE J Sel Areas Commun 33(1):1–13

26. Wu Y, Wang J, Qian L, Schober R (2015) Optimal power control for energy efficient D2D communication and its distributed implementation. IEEE Commun Lett 19(5):815–818

27. Erturk M, Mukherjee S, Ishii H et al (2013) Distributions of transmit power and SINR in device-to-device networks. IEEE Commun Lett 17(2):273–276

28. Fodor G, Belleschi D, Johansson M, Abrardo A (2013) A comparative study of power control approaches for device-to-device communications. In: IEEE international conference on communications

29. Sheng M, Liu J, Zhang Y, Sun H, Li J (2015) On transmission capacity region of D2D integrated cellular networks with interference management. IEEE Trans Commun 63(4):1383–1399

30. Lei L, Zhang Y, Shen X, Lin C, Zhong Z (2013) Performance analysis of device-to-device communications with dynamic interference using stochastic Petri nets. IEEE Trans Wirel Commun 12(12):6121–6141

31. Xu W, Liang L, Zhang H, Jin S, Li J, Lei M (2012) Performance enhanced transmission in device-to-device communications: beamforming or interference cancellation. In: IEEE global communications conference, Anaheim, California, USA

32. Min H, Lee J, Park S, Hong D (2011) Capacity enhancement using an interference limited area for device-to-device uplink underlaying cellular networks. IEEE Trans Wirel Commun 10(12):3995–4000

33. Golrezaei N, Dimakis A, Molisch AF (2014) Scaling behavior for device-to-device communications with distributed caching. IEEE Trans Inf Theory 60(7):4286–4298

34. Li Y, Wang W (2014) Message dissemination in intermittently connected D2D communication networks. IEEE Trans Wirel Commun 13(7):3978–3990

35. Ji M, Caire G, Molisch AF (2013) Fundamental limits of distributed caching in D2D wireless networks. In: IEEE Information Theory Workshop, Seville

36. Borst S, Gupta V, Walid A (2010) Distributed caching algorithms for content distribution networks. In: IEEE international conference on computer communications, San Diego, CA/USA, Sep

37. Antonopoulos A, Kartsakli E, Verikoukis C (2014) Game theoretic D2D content dissemination in 4G cellular networks. IEEE Commun Mag 52(6):125–132

38. Hu J, Yang L, Hanzo L (2013) Mobile social networking aided content dissemination in heterogeneous networks. China Commun 10(6):1–13
39. Zhou L, Chao H (2011) Multimedia traffic security architecture for internet of things. IEEE Netw 25(3):35–40
40. Zhou L, Zhang Y, Song K, Jing W, Vasilakos A (2011) Distributed media-service scheme for P2P-based vehicular networks. IEEE Trans Veh Technol 60(2):692–703
41. Ioannidis S, Chaintreau A, Massoulie L (2009) Optimal and scalable distribution of content updates over a mobile social network. In: IEEE international conference on computer communications
42. Shakkottai S, Johari R (2010) Demand-aware content distribution on the internet. IEEE/ACM Trans Netw 18(2):476–489
43. Bass F M (1969) A new product growth model for consumer durables. Manag Sci 15:215–227
44. Zhao, J, Cao G (2008) VADD: vehicle-assisted data delivery in vehicular ad hoc networks. IEEE Trans Veh Technol 57(3):1910–1922
45. Zhao J, Zhang Y, Cao G (2007) Data pouring and buffering on the road: a new data dissemination paradigm for vehicular ad hoc networks. IEEE Trans Veh Technol 56(6):3266–3277
46. Alam M, Yang D, Rodriguez J, Abd-Alhameed R (2014) Secure device-to-device communication in LTE-A. IEEE Commun Mag 52(4):66–73
47. Panaousis E, Alpcan T, Fereidooni H, Conti M (2014) Secure message delivery games for device-to-device communications. Springer, Cham, pp 195–215
48. Nobach L, Hausheer D (2014) Towards decentralized, energy and privacy-aware device-to-device content delivery. In: IFIP international federation for information processing, pp 128–132
49. Kwon H, Hahn C, Kim D, Kang K, Hur J (2014) Secure device-to-device authentication in mobile multi-hop networks. Springer, Cham, pp 267–278
50. Yue J, Ma C, Yu H, Zhou W (2013) Secrecy-based access control for device-to-device communication underlaying cellular networks. IEEE Commun Lett 17(11):2068–2071
51. Zhang H, Wang T, Song L, and Han Z (2014) Radio resource allocation for physical-layer security in D2D underlay communications. In: IEEE international conference on communications, Sydney, Australia, June
52. Zhu D, Swindlehurst A, Fakoorian S, Xu, W, Zhao C (2014) Device-to-device communications: the physical-layer security advantage. In: IEEE international conference on acoustic, speech and signal processing, Florence, Italy, May
53. Yu W, Liu K (2007) Game theoretic analysis of cooperation stimulation and security in autonomous mobile ad hoc networks. IEEE Trans Mob Comput 6(5):507–521
54. Niu B, Zhao H, Jiang H (2011) A cooperation stimulation strategy in wireless multicast networks. IEEE Trans Signal Process 59(5):2355–2369
55. Mahmoud M, Shen X (2012) FESCIM: fair, efficient, and secure cooperation incentive mechanism for multihop cellular networks. IEEE Trans Mob Comput 11(5):753–766
56. Kang X, Wu Y (2013) A game-theoretic approach for cooperation stimulation in peer-to-peer streaming networks. IEEE international conference on communications
57. Zhou H, Chen J, Fan J, Du Y, Das S (2013) ConSub: incentive-based content subscribing in selfish opportunistic mobile networks. IEEE J Sel Area Commun Supplement 31(9):669–679
58. Zhang G, Yang K, Liu P, Yang X, Ding E (2012) Resource-exchange based cooperation stimulating mechanism for wireless ad hoc networks. In: IEEE international conference on communications
59. Zhang G, Cong L, Ding E, Yang K, Yang X (2011) Fair and efficient resource sharing for selfish cooperative communication networks using cooperative game theory. In: IEEE international conference on communications
60. Chen X, Proulx B, Gong X, Zhang J (2013) Social trust and social reciprocity based cooperative D2D communications. In: ACM International Symposium on Mobile Ad Hoc Networking and Computing (MOBIHOC), Bangalore, India
61. Chen X, Proulx B, Gong X, Zhang J (2015) Exploiting social ties for cooperative D2D communications: a mobile social networking case. IEEE/ACM Trans Netw 23(5):1471–1484

62. 3GPP, TR 23.703, v. 0.4.0, Study on architecture enhancements to support proximity services (ProSe). Release 12, June 2013
63. 3GPP, TR 33.401, v. 12.9.0 security architecture. Release 12, Sept. 2013
64. Zhang A, Chen J, Zhou L (2015) Content dissemination and security in device-to-device (D2D) communication. Future of wireless networks: architectures, protocols, and services. Springer, New York
65. Li Z, Shen H (2012) Game-Theoretic analysis of Cooperation incentive strategies in mobile Ad Hoc networks. IEEE Trans Mob Comput 11(8):1287–1303
66. Chen T, Zhu L, Wu F, Zhong S (2011) Stimulating cooperation in vehicular ad hoc networks: a coalitional game theoretic approach. IEEE Trans Veh Technol 60(2):566–579

# Chapter 2
# Secure Data Transmission Protocol

**Abstract** In Device-to-Device (D2D) communications, data transmission is a very important application. However, due to the fact that the connections happen directly between the proximity devices, they may subject to variety security threats such as data modification and fabrication, privacy violation. This chapter proposes a secure data sharing protocol through a cryptography approach. Specifically, digital signature combing with mutual authentication between the evolved NodeB (eNB) and users guarantees the entity authentication, data authority and integrity, transmission non-repudiation as well as traceability. Meanwhile, data confidentiality is achieved through the symmetric encryption. The key hint transmission between the eNB and user equipments (UEs) realizes reception non-repudiation. Free-riding attack is detected by keeping a record of the data sharing behaviors for the UEs in the eNB thus the system availability is improved.

This chapter is organized as follows: Network architecture and preliminaries are given in Sect. 2.1. In Sect. 2.2, the proposed protocol is presented in detail. Section 2.3 analyzes its performance including security properties and overhead. In Sect. 2.4, the scheme is discussed by proposing a modified protocol. Finally, Sect. 2.5 concludes this work. Note that the majority of the contents of this chapter are based on our previous work [1].

## 2.1 System Model

### 2.1.1 Network Architecture

Generally, a D2D communication system includes four parts: GateWay (GW), eNB, UEs, and service provider (SP), as shown in Fig. 2.1 [1].

**GW** The GW works as the gate from the local network to the core network and routes IP packets from/to the Internet. It can detect the potential D2D traffic with Proximity Service Control Function (PSCF). The PSCF earmarks the traffic flows and searches pairs of D2D enabled devices.

© The Author(s) 2016
A. Zhang et al., *Security-Aware Device-to-Device Communications Underlaying Cellular Networks*, SpringerBriefs in Electrical and Computer Engineering, DOI 10.1007/978-3-319-32458-6_2

**Fig. 2.1** System model for a
general D2D communication
scenario

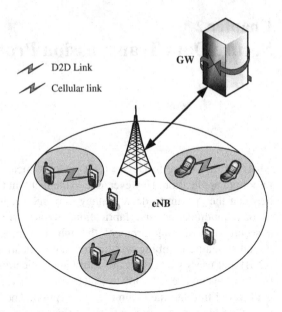

eNB  The eNB connects the mobile phone to the GW. It is an important infrastructure in Evolved Universal Terrestrial Radio Access (E-UTRA) of LTE-A network.
The main roles of eNB include: (1) resource allocation and coordination of devices
by ensuring two peers meet in space, time, and frequency, (2) power control for the
transmitter of the cellular users to limit the interference, and (3) user authentication
in cellular network. Additionally, the eNB also works as the trust authority. It usually
has a high storage capacity and strong computational capacity.

UEs  UEs are the D2D communication terminals. The UEs may share information
by D2D communication without increasing additional traffic to the eNB. In the
book, we assume that all the UEs are served by the same eNB. Note that in LTE-A
network, UE is designed with the function of mutual authentication with the eNB.

SP  SP provides authentic content when the system is established. The contents are
firstly sent to partial UEs so that they can then share the material with other devices
through D2D communications. The SP may quit the system when most of the UEs
obtain the contents.

In the system, we assume that the eNB and GW are completely trustable, i.e.,
they can't be comprised by the attackers. Moreover, the SP is honest enough to
provide the correct and authentic source content while the UEs may be comprised
of or captured by some adversaries.

## 2.1.2 Preliminaries

**Bilinear Pairing [2]** Let $\mathbb{G}_1$ and $\mathbb{G}_2$ be two multiplicative cyclic groups of the same prime order $q$. Let $g_1$ and $g_2$ be two generators of $\mathbb{G}_1$ and $\mathbb{G}_2$, respectively. A mapping $\hat{e} : \mathbb{G}_1 \times \mathbb{G}_1 \to \mathbb{G}_2$ is called an admissible bilinear map if it satisfies the following properties:

1. Bilinear: For all $W, Q \in \mathbb{G}_1$ and $a, b \in \mathbb{Z}_q^*$, we have $\hat{e}(W^a, Q^b) = \hat{e}(W, Q)^{ab}$.
2. Symmetric: $\hat{e}(W, Q) = \hat{e}(Q, W)$.
3. Non-degenerate: $\hat{e}(W, Q) \neq 1_{\mathbb{G}_2}$, where $W, Q \neq 1_{\mathbb{G}_1}$.
4. Computable: $\hat{e}$ is efficiently computable.

As mentioned in [3], such an admissible map can be constructed by the modified Weil of Tate pairing on elliptic curve and a 160-bit prime order $q$ is assumed to reach 80-bit security level.

**Definition 2.1 (Bilinear Parameter Generator *Gen*).** A bilinear parameter generator *Gen* is a probabilistic algorithm that takes a security parameter $k$ as input and outputs a tuple $(q, g_1, g_2, \mathbb{G}_1, \mathbb{G}_2, \hat{e})$, where $\mathbb{G}_1$ and $\mathbb{G}_2$ are two multiplicative cyclic groups of order $q$. $g_1$ and $g_2$ are two generators of $\mathbb{G}_1$ and $\mathbb{G}_2$, and $\hat{e} : \mathbb{G}_1 \times \mathbb{G}_1 \to \mathbb{G}_2$ is an admissible bilinear map.

**Discrete Logarithm Problem (DLH)** DLH is of fundamental importance for many practical public-key algorithms.

**Definition 2.2 (Discrete Logarithm Problem (DLH) in $\mathbb{Z}_q^*$ [4]).** Given the finite cyclic group $\mathbb{Z}_q^*$ of order $q$ and a primitive element $\alpha \in \mathbb{Z}_q$ and another element $\beta \in \mathbb{Z}_q$, the DLP is the problem of determining the integer $1 \leq x \leq q$ such that

$$\alpha^x = \beta \mod q$$

It is assumed to be intractable within polynomial time to solve the DLP problem [4].

Diffie–Hellman Key Exchange (DHKE) [5], which is based on DLH, provides a practical solution to the key distribution problem, i.e., it enables two parties to derive a common secret key by communicating over an insecure channel. The basic idea underlying DHKE is that the exponentiation in $\mathbb{Z}_q^*$ is a one-way function and commutative, i.e.,

$$key = (x^a)^b \equiv (x^b)^a \mod q.$$

Here, $A = x^a$ and $B = x^b$ are named key hints. By exchanging their key hints, the two parties are able to compute their shared keys, respectively. We refer to [5] for more detailed descriptions.

**Elliptic Curve Cryptosystems (ECC)** An elliptic curve is a special type of polynomial equation, defined as following:

**Definition 2.3 (Elliptic Curve).** The elliptic curve over $\mathbb{Z}_q, q > 3$, is the set of all pairs $(x, y) \in \mathbb{Z}_q$, which fulfill

$$y^2 = x^3 + ax + b \mod q$$

together with an imaginary point of infinity $\infty$, where $a, b \in \mathbb{Z}_q$ and the condition $4a^3 + 27b^2 \neq 0 \mod q$.

As presented in [4], the points on an elliptic curve together with $\infty$ have cyclic subgroups. Under certain conditions all points on an elliptic curve form a cyclic group. Let $\sharp E$ denotes the number of points on the curve. The Elliptic Curved Discrete Logarithm Problem (ECDLP) is defined as following:

**Definition 2.4 (Elliptic Curved Discrete Logarithm Problem (ECDLP) [4]).** Given an elliptic curve $E$, we consider a primitive element $P$ and another element $T$. The ECDLP problem is finding the integer $d$, where $1 \leq d \leq \sharp E$, such that:

$$\underbrace{P + P + \cdots + P}_{d} = dP = T$$

It is assumed to be intractable within polynomial time to solve the ECDLP problem [4]. In cryptosystems, $d$ is the private key which is an integer, while the public key $T$ is a point on the curve with coordinates $T = (x_T, y_T)$.

## 2.2   Secure Data Transmission Protocol

In the proposed protocol, the system is initialized before data transmission. During the processes of data transmission between the UEs, the entity eNB and GW are involved to guarantee security. For clarity of presentation, the notations used throughout the paper are listed in Table 2.1.

### 2.2.1   System Initialization

System initialization aims to generate the system parameters and register the users. Specifically, it includes four steps as shown in Fig. 2.2.

*Step 1 System Parameter Generation* The eNB generates the tuple $(q, g_1, g_2, \mathbb{G}_1, \mathbb{G}_2, \hat{e})$ by running $Gen(k)$, where $k$ is the security parameter. Then a secure symmetric encryption function $Enc_s()$ and two hash functions $H_1$ and $H_2$ are selected by the eNB for the system. Here $H_1 : \{0, 1\}^* \rightarrow \mathbb{Z}_q^*$, $H_2 : \{0, 1\}^* \rightarrow \mathbb{G}_2$. Then, $para = (q, g_1, g_2, \mathbb{G}_1, \mathbb{G}_2, \hat{e}, Enc_s(), H_1, H_2)$ are published as the system parameters.

**Table 2.1** Notations and description

| Notation | Description | Notation | Description |
|---|---|---|---|
| TA | Trust authority | $X_i$ | The public key of $UE_i$ or $SP$ |
| SP | Service provider | $x_i$ | The private key of $UE_i$ or $SP$ |
| $UE_i$ | The $i$th user equipment | $H_1()$ | A secure hash function such as $\{0,1\}^* \rightarrow \mathbb{Z}_q^*$ |
| $RID_i$ | Real identity of $UE_i$ or $SP$ | $H_2()$ | A secure hash function such as $\{0,1\}^* \rightarrow \mathbb{G}_2$ |
| $PID_i$ | Pseudo identity of $UE_i$ or $SP$ | $Enc_s()$ | Symmetric encryption algorithm with key $s$ |
| $Sgn_s()$ | Signature function with key $s$ | $Dec_s()$ | Symmetric decryption algorithm with key $s$ |
| $P_i$ | The portion index of data | $a\|b$ | String concatenation of $a$ and $b$ |

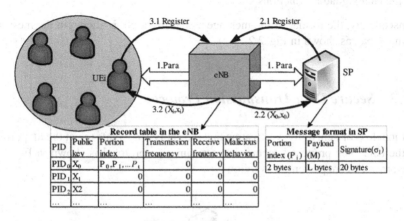

**Fig. 2.2** System initialization

*Step 2 SP Registration* Upon receiving the registration request from the SP, the eNB firstly computes $PID_0 = H_1(RID_0)$, where $PID_0$ and $RID_0$ are the pseudo identity and real identity for the SP, respectively. Then, it randomly selects an integer $x_0 \in \mathbb{Z}_q^*$ and computes $X_0 = g^{x_0}$. $X_0$ and $x_0$ perform as the public key and the private key to the SP. Finally, the eNB sends the public/private key $(X_0, x_0)$ pair to the SP via a secure channel.

*Step 3 UE Registration* Similar to *Step 2*, when $UE_i$ with real identity $RID_i$ registers to the eNB, it computes $PID_i = H_1(RID_i)$ as the pseudo identity for the UE. Then it randomly selects an integer $x_i \in \mathbb{Z}_q^*$ and computes $X_i = g^{x_i}$. The eNB sends public/private key $(X_i, x_i)$ to the $UE_i$ through a secure channel. Additionally, the tuple $(X_0, PID_0)$ is also sent to the register.

*Step 4 System Setup* The eNB records the entities' status and their data sharing behavior, as shown in the record table of the eNB in Fig. 2.2. Due to the fact the data may be too large (e.g., video clips), they may be divided into several frames.

Thus, "Portion index $(P_i)$" is introduced in the system to index the data.[1] The item "Transmission frequency" counts the amounts that the device transmits data to their neighbors while "Receiving frequency" denotes the receiving amount. "Malicious behavior amount" tracks the times the entity transmits fake or fabricated message. All the initial values are "0". The eNB also stores the original data $M$ and the corresponding portion index for the verification of the message.

Importantly, the data is signed by the SP in order to guarantee data authority and integrity. The signature process is completed offline to reduce the data transmission latency. Specifically, the SP computes the signature $\sigma_0 = Sgn_{x_0}(M)$.

*Remark 2.1.* The function $Sgn_s(M)$ is a signature on $M$ with key $s$. Due to the mutual authentication in cellular mode, the signature formation may be simply realized by a secure hash function, i.e., $Sgn_s(M) = H_2(M)^s$. It is the same case for the later signature functions.

Consequently, the format of the message stored in the SP is shown in the message format table, as shown in Fig. 2.2.

### 2.2.2   Secure Data Transmission Protocol

Without loss of generality, we assume that the entity $UE_i$ wants to get the $i$th portion of the data. The proposed secure data transmission protocol is shown in Fig. 2.3. The processes are presented as follows:

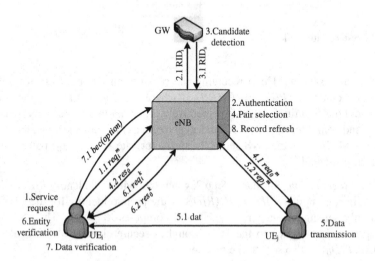

**Fig. 2.3** Secure data sharing protocol

---

[1]We assume that the data is divided into $t$ portions, thus $P_i$ may be $0, 1, 2, \cdots t$. Note that "0" represents the entity obtaining no share.

*Step 1 Service Request* In order to generate the communication key $k_c$, $UE_i$ randomly selects $a \in \mathbb{Z}_q^*$ and computes the key hint $A = g^a$. Additionally, HMAC is introduced to realize integrity and authentication. The HMAC is specified as internet standard RFC 2104 in the protocol, i.e., the HMAC for message $m$ is

$$HMAC_k(m) = h[(k^+ \oplus opad) \| h[(k^+ \oplus ipad) \| m]], \qquad (2.1)$$

where $k^+$ is the key padded out to size, and $opad = 00110110, 00110110, \ldots,$ $00110110$, $ipad = 01011100, 01011100, \ldots 01011100$ are specified padding constants, $h(\cdot)$ is a cryptographic hash functions such as SHA-1.

Thus, the service request message is formulated as $req_i^m = (PID_i \| z \| P_i \| h[(x_i^+ \oplus opad) \| h[(x_i^+ \oplus ipad) \| PID_i \| z \| P_i]])$ and sent to the eNB by $UE_i$, where $P_i$ denotes the expected portion index.

It is worthy noting that $h[(x_i^+ \oplus opad) \| h[(x_i^+ \oplus ipad) \| PID_i \| z \| P_i]]$ is denoted by $h(\bullet, x_i)$, where $\bullet$ denotes the message attached by the HMAC for the simplification of expression. In the next steps, all the HMACs are expressed as the same formation.

*Step 2 Authentication* When receiving the request message, the eNB verifies its integrity and verification,[2] authenticates the requester in the normal cellular communication mode, and extracts the real identity $RID_i$. Then it checks if $RIDi$ is in the record table. If not, the message is ignored. Otherwise, the eNB will forward $RIDi$ to the GW for peer discovery and finding out the potential transmitters.

*Step 3 Candidate Detection* The GW performs peer discovery and searches the potential D2D pairs for $UE_i$. Then it responds the eNB with the real identities (RIDs) of candidates.

*Step 4 Pair Selection* The devices, the portion index of which meets with the demand, are selected as the candidates. However, the device, denoted by $UE_j$, holding the minus share frequency, is selected to act as the transmitter for the balance of load and fairness. Then the eNB randomly selects $u \in \mathbb{Z}_q^*$ and computes $U = g^u$ as a key hint. Then the request message $req_0^m = (PID_j \| PID_i \| z \| u \| P_i \| h(\bullet, x_j))$ is then sent to $UE_j$ by the eNB. Simultaneously, the eNB responds $UE_i$ with message $res_0^m = (PID_i \| PID_j \| X_j \| P_i \| h(\bullet, x_0))$.

Step 5 Data Transmission After receiving a communication request message $req_0^m$, $UE_j$ randomly selects $b \in \mathbb{Z}_q^*$, and computes the key hint $B = g^b$ as well as the communication key $k_c = A^b = g^{ab}$. Then $UE_j$ encrypts the message $M$ by computing $M' = Enc_{k_c}(M)$ with $k_c$ signs the message by computing $\sigma_1 = Sgn_{x_j}(M')$ with its private key $x_j$.

Therefore, the data is shaped in the format as $dat = (PID_i \| PID_j \| P_i \| M' \| T_s \| \sigma_0 \| \sigma_1)$, as shown in Table 2.2 and sent to $UE_i$. Note that signature $\sigma_0$ is finished offline by the SP and the timestamp $T_s$ is adopted to resist the replay attack.

---

[2]In the next steps, all the messages will be checked in the same. We omit this procedure in the other steps for shortening the space.

**Table 2.2** Data format during transmission

| $PID_i$ | $PID_j$ | Portion index ($P_i$) | $Enc(M)$ ($M'$) | Timestamp ($T_s$) | Signature ($\sigma_0$) | Signature ($\sigma_1$) |
|---------|---------|-----------------------|-----------------|-------------------|------------------------|------------------------|
| 2 bytes | 2 bytes | 2 bytes               | $L$ bytes       | 2 bytes           | 20 bytes               | 20 bytes               |

It is worthy noting that the key hint $B$ is not included in the data $dat$. This is because the key hint is sent to the eNB to realize receiving non-repudiation. Specifically, $UE_j$ randomly selects $v \in \mathbb{Z}_q^*$ and computes $V = g^v$. The shared key $k_s$ between the eNB and $UE_j$ is $k_s = U^v$. Then the key hint $B$ is encrypted with $k_s$, $B' = Enc_{k_s}(B)$. As a result, a report $rep_j^m = (PID_i\|PID_j\|P_i\|B'\|v\|T_s\|h(\bullet, x_j))$ is sent to the eNB for the data transmitting event so that its record may be refreshed.

*Step 6 Entity Verification* The receiver $UE_i$ extracts $PID_j$ from the message $dat$ and checks its validity with the pseudo identity obtained from the eNB. If not, the packet is dropped. Otherwise, it checks the validity of the message by verifying signature $\sigma_1$. If it is valid, the data is considered to be sent by $UE_j$. Then, $UE_i$ sends a key hint request message $req_i^k = (PID_i\|PID_j\|P_i\|T_s\|h(\bullet, x_i))$ to the eNB for decrypting the message $M'$,

After receiving $req_i^k$, the eNB first checks if the time information of the message is in the time allowable window. If so, it decrypts $B'$ with $k_s = V^u$ and responds with $res_0^k = (PID_i\|PID_j\|P_i\|y\|T_s\|T_i\|h(\bullet, x_i))$, where timestamp $T_i$ is used to record the feedback time as analyzed in *Step 8*.

*Step 7 Data Verification* $UE_i$ first computes $k_c = B^a = g^{ab}$ with the reception of key hint $B$. Thus the original message $M$ is revealed by decrypting $M'$. Then the authority of the data is checked by verifying signature $\sigma_0$. If it is valid, the message is recognized as authorized. Otherwise, the impersonation attack may have occurred. In this case, $UE_i$ will report a beacon

$$bec = (PID_i\|PID_j\|P_i\|M'\|T_s\|\sigma_0\|\sigma_1\|h(\bullet, x_i)),$$

to the eNB within the timestamp $T_i'$, which satisfies that $T_i' < T_i + \Delta T$ ($\Delta T$ is the predefined time scale).

*Step 8 Record Refresh* If any feedback beacon arrives during the waiting time scale, the eNB first checks the validity of $\sigma_0$ as $UE_i$ had implemented. Note that the original message $M$ is not obtained by deciphering $M'$ because the eNB doesn't know $k_c$. Instead, it refers to its storage for $M$. If the signature $\sigma_0$ is invalid indeed, the record in eNB is refreshed as follows:

1. The eNB verifies the validity of signature $\sigma_1$ to ensure the fake message is sent by $UE_j$,
2. The malicious behavior amount record of $UE_j$ adds one.

If no feedback beacon arrives during the waiting time scale or the signature $\sigma_0$ is valid, the record table is refreshed as

1. $P_i$ is inserted in the portion index item of $PID_i$,
2. the transmission frequency of $PID_j$ adds one and,
3. the receiving frequency of $PID_i$ adds one.

## 2.3 Performance Analysis

The proposed protocol is designed to migrate a variety of potential threats and meet with the security requirements. This section analyzes its security properties and overhead.

### 2.3.1 Security Properties

*Data Confidentiality and Integrity* Data confidentiality is guaranteed by encryption. As is known to all, the security of encryption algorithm is determined by the security of the secret key. In the proposed protocol, the communication key shared between the transmitter and receiver is generated by DHKE based on DLP. The eavesdropper, who even gets both key hints $g^a$ and $g^b$, still can't derive the shared key $g^{ab}$ under the DLP assumption. Thus the key is secure and the encrypted data can't be revealed to the eavesdroppers. Meanwhile, HMAC is applied to provide message integrity. Furthermore, in order to protect data authority, the signature of the SP $\sigma_0$ is added in the data *dat*. Therefore, the data integrity and authority is guaranteed with the verification of signature $\sigma_0$.

*Entity Authentication* Entity authentication takes places between the eNB and UE as well as between the UEs in the D2D communications. Firstly, the eNB and UE authenticate each other by the normal cellular communication. Then, the eNB refers to its record table to check the membership of the requesting UE. As for the D2D communication, entity authentication is implemented by the verification of the signature $\sigma_1$.

*Conditional Privacy Preservation* The UEs communicate with each other using pseudo identity, which is the secure one-way hash value of the real identity. It is computationally hard to identify the real identity of the entity. Therefore, the privacy of the UEs is preserved. However, their real identities are disclosed to the eNB thus the privacy preservation property is conditional.

*Resistant to Free-Riding Attacks* Free-riding attack is resisted technically by keeping a record table in the eNB. Specifically, free riders are easily revealed since their transmission frequency is low while their receiving frequency is high. The free-riders may be punished by being not delivered any message any more or excluded from the membership.

*Non-repudiation* The proposed protocol provides both transmission non-repudiation and receiving non-repudiation. Specifically, the transmitter cannot deny the event of data transmission because of its signature $\sigma_1$. Note that it also offers the evidence of data transmitting behavior, which may be proofed to be meritorious or malicious by the verification of signature $\sigma_0$.

On the other hand, the key hint request message provides no opportunity for the receiver to deny its data receiving event. Notably, there may be irrational attacker who only receives data but not sending a key hint request. In this case, the receiver (attacker) cannot get the original message and the eNB will not refresh its record. Thus, the transmitter becomes the victim who does transmit the data but its transmission frequency is not increased. One solution to this problem might introduce cooperation stimulation such as reputation mechanism into the system, which is an open issue.

### 2.3.2 Overhead

In this section, the storage overhead, computational overhead, and communication overhead are analyzed based on the following simulation settings: The CPU processing speed is set to be 1 GHz as in most intelligent mobile terminals. Meanwhile, for a typical 80-bit security level, the system parameters $k$ and $q$ are 80 bits and 160 bits, respectively, and Advanced Encryption Standard (AES) serves as the symmetric encryption algorithm.

*Storage Overhead* Every UE stores a public/private key pair and a pseudo identity, which occupy $20 + 2 + 2 = 24$ bytes storage overhead. In addition, the random number, which is selected for generating key hint, is also stored at the cost of 2 bytes. Suppose the original material ($L$ bytes) is divided into $t$ portions. As for every portion of the material, an additional overhead due to signature and portion index yield $20 + 2$ bytes for each portion. Thus the total storage overhead caused by the data is $L + 22t$ bytes. As a result, in each UE, $L + 22t + 26$ bytes storage overhead is needed. Figure 2.4 shows that the storage overhead in the proposed protocol linearly increases with $t$. As $L$ increases, the total storage overhead increases. Generally, the size of the message dominates the other two items.

*Computational Overhead* As the pairing, multiplication, and encryption operation dominate the computational overhead, we only consider the three operations in the following estimation. In [6], the algorithm runs on an Intel Pentium 3.0 GHz processor for an MNT curve of embedding degree $= 6$ and 160-bit $q$. It takes 4.5 ms for one pairing operation and 0.6 ms for one multiplication in $\mathbb{G}$. Our implementation is executed on the similar settings except the 1 GHz microprocessor. Thereafter, the computational overhead for one pairing and point multiplication of ours can be roughly estimated as $\frac{4.5 \times 3}{1} = 13.5$ ms and $\frac{0.6 \times 3}{1} = 1.8$ ms, respectively. It is reported in [7] that it takes 0.984 μs for AES to encrypt 64 bytes data on a

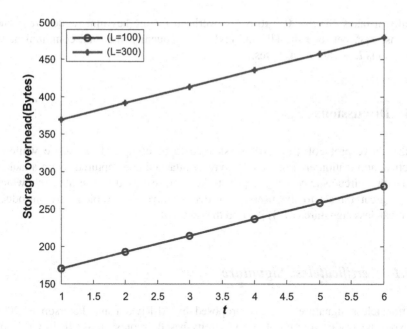

**Fig. 2.4** Storage overhead varies with $t$

1.8 GHz laptop PC thus the encryption may take $0.984 \times 1.8 = 1.771 \, \mu s$ on a 1 GHz microprocessor. Then, our processor may take $\frac{1.771}{64} L = 0.028L \, \mu s$ for encrypting an $L$ bytes data.

In the proposed protocol, it requires two exponential computations and one encryption operation for the transmitter ($UE_j$) to formate the data in *Step 5*. This process introduces $2t_e + t_s$ time cost. The receiver UE ($UE_i$) spends $4t_p$ to authenticate the validity of signature $\sigma_0$ and $\sigma_1$. Therefore, the total computational overhead of the users is $4t_p + 2t_e + t_s$ time cost. It is worthy noting that the eNB also performs computation to decrypt the key hint, which yields $2t'_e + t'_s$ time cost, where $t'_n$ and $t'_s$ are time cost in eNB for one exponential computation and decryption, respectively.

*Communication Overhead* Due to the fact the overhead of application message dominates the communication overhead in most cases [8], we only consider the application layer overhead for simplification. The communication overhead for service request and response between the eNB and UE is 44+46=90 bytes, where the first term represents the length of $req_i^m$, the second one represents the overhead of $res_0^m$. It takes $2 + 2n_c$ communication overhead for the candidate detection between the eNB and GW, where $n_c$ is the number of the candidate detected by GW. Additionally, the eNB sends a request message $req_i^m$ to the selected candidate for informing the transmitter, causing 66 bytes overhead. From Table 2.2, it can be found that the device to device data transmitting cost $L + 48$ bytes overhead.

Finally, it takes 68+28+50=146 bytes overhead for the key hint request $req_i^k$ and response $res_0^k$ between the UE and eNB. Consequently, the total communication overhead is $L + 186 + 2n_c$ bytes.

## 2.4  Discussions

In the above protocol, the eNB is assumed to be completely trustable while in practical application environment, it may be attacked and captured by an attacker. Under this circumstance, all the private keys are exposed to the attacker, which brings great threats to the users. In order to solve the problem, we introduce certificateless signature (CLS) [9] into the protocol.

### 2.4.1  Certificateless Signature

Certificateless signature is firstly proposed by Al-Riyami and Paterson in 2003 to avoid the key escrow problem in identity-based cryptosystem [9]. In CLS, the users' private key is not generated by the Key Generator Center (KGC) alone but a combination of the contributions of the KGC and the user. The KGC has no information about the user's private key but can authenticate its public key. The key escrow problem is solved technically in this way.

Generally, a certificateless signature scheme consists of the following six algorithm [10, 11]:

- **Setup**. This algorithm is performed by the KGC. It takes as input a security parameter $k$, and outputs a master private key $x$, a corresponding master public key $P_{pub}$, and public parameters $params$. The KGC publishes $params$ and keeps the master key $x$ as secret.
- **Partial-private-key-extract**. This algorithm is performed by the KGC. It takes as input $params$, $x$, and identity $ID$ of a user, and outputs a partial-private-key $s_{ID}$.
- **Set-private-key**. This algorithm is performed by the user. It takes as input the parameter $params$, the identity $ID$ of the user, and the partial-private-key. Then, it returns the user's private key $sk_{ID}$.
- **Set-public-key**. This algorithm is performed by user. It takes as input the parameters params, user identity ID, and the user's private key. It returns his public key $pk_{ID}$.
- **Sign**. This algorithm is performed by the signer. It takes as input the parameter $params$, the message $m$, signer's $ID$, and private key $sk_{ID}$. It outputs a signature $\sigma$ on $m$.
- **Verify**. This algorithm is performed by the verifier. It takes as input the signature $\sigma$, message $m$, an identity $ID$, and the corresponding public key $pk_{ID}$. It returns 1 if $\sigma$ is valid and 0 otherwise.

**Fig. 2.5** System initialization of CLS-based protocol

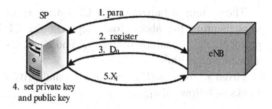

To avoid reinventing the wheel, we refer to [9, 12, 13] for a more detailed introduction of CLS. In this brief, we mainly describe how to apply CLS in the protocol to enhance the system security. For simplification of presentation, the above protocol is named ID-based protocol, and the protocol using CLS is named CLS-based protocol.

### 2.4.2 CLS-Based Data Transmission Protocol

The differences between the two protocols lie in system initialization, as shown in Fig. 2.5. In the CLS-based system, the eNB works as the KGC and the system is initialized as follows:

*Step 1 System Parameter Generation* Except the parameters published in the ID-based protocol, the eNB additionally completes the following tasks:

1. Choose an arbitrary generator $P \in \mathbb{G}_1$.
2. Select $r \in \mathbb{Z}_q^*$ at random as the master key of the system and computes $P_0 = rP$ as the public key of the system.

Then, the system parameters $para = (q, g_1, g_2, \mathbb{G}_1, \mathbb{G}_2, \hat{e}, P_0, Enc_s(), H_1, H_2)$ are published.

*Step 2 SP Registration* When the SP with real identity $RID_O$ registers to the eNB, the eNB performs algorithm **Partial-private-key-extract** by computing $Q_0 = PID_0 = H_2(RID_0)$ and setting $D_0 = s * Q_0$ as the partial-private-key. Note that the SP can verify the correctness of the partial-private-key by checking $\hat{e}(D_0, P) = \hat{e}(Q_0, P_0)$. The eNB sends the partial-private-key to the SP through a secure channel.

The SP executes algorithm **Set-private-key** by randomly choosing $w \in \mathbb{Z}_q^*$ and computes $x_0 = wQ_0 + D_0$ as its secret key. Then the SP carries out **Set-public-key** by computing $X_0 = x_0 * P$ as its public key. In order to authenticate itself to the eNB, the SP computes $Y_0 = D_0 * P$ to the eNB with its public key, i.e., the SP sends $(X_0, Y_0)$ to the eNB. After receiving the tuple $(X_0, Y_0)$, the eNB checks $Y_0 = D_0 * P$ and then fills in the table with the public key $X_0$ at the item of $PID_0$.

*Step 3 UE Registration* It is similar to the SP registration. At the end of this process, the UE ($UE_i$) keeps the private key $x_i$ and sends the public key $X_i$ to the eNB for record.

*Step 4 System Setup* It is identical to the ID-based protocol.

The system initialization for CLS-base secure data transmission is finished by implementing the above four steps. In the CLS-base secure data transmission protocol, the signature of the SP and the UEs is substituted by the following signature algorithm.

Given *params*, $RID_0$, message $m$, and private key $x_0$, the signature algorithm works as follows [14]:.

1. Compute $Q_0 = H_2(RID_0)$.
2. Choose a random value $f \in Z_q^*$ and set $U = fQ_0$.
3. Set $h = H_1(m||U)$.
4. Compute $V = (f + h)x_0$.
5. Set $\sigma = (U, V)$ as the signature of $m$.

Then, the verification algorithm is replaced by the following algorithm correspondingly.

Given signature $\sigma$, $RID_0$, message $m$, and public key $X_0$, the verification algorithm works as follows:

1. Compute $Q_0 = H_2(RID_0)$.
2. Compute $h = H_1(m||U)$.
3. Check whether $e(P, V) = e(P_0 + X_0, U + hQ_0)$. If not, then reject the signature else accept it.

*Remark 2.2.* In this brief, we just give an example for the CL-PKC-based protocol. Actually, there have been abundant researches on CL-PKC and different CL-PKC schemes have been proposed [9, 12, 13]. In practical application environment, we may choose proper CL-PKC mechanism for the secure data transmitting protocol.

## 2.5   Conclusions

In this chapter, a secure data sharing protocol is proposed for D2D communication underlaying cellular network. The proposed protocol is designed based on digital signature and symmetric encryption to achieve security goals in terms of data confidentiality and integrity, transmission and reception non-repudiation, entity authentication, and free-riding resistant. Moreover, a CL-PKC-based secure data sharing protocol is proposed in this chapter for the circumstance that the eNB may be captured by attackers.

## References

1. Zhang A, Chen J, Hu R, Qian Y (2015) SeDS: Secure data sharing strategy for D2D communication in LTE-Advanced networks. IEEE Trans Veh Technol. doi: 10.1109/TVT.2015.2416002
2. Boneh D, Franklin M (2001) Identity-based encryption from the Weil pairing. In: International cryptology conference. LNCS, vol 2139. Springer, Berlin, pp 213–229

3. Miyaji A, Nakabayashi M, Takano S (2001) New explicit conditions of elliptic curve traces for FR-reduction. IEICE Trans Fundam E84-A(5):1234–1243
4. Paar C, Pelzl J (2010) Understanding cryptography - a textbook for students and practitioners. Springer, New York
5. Diffie W, Hellman ME (1976) New directions in cryptography. IEEE Trans Inf Theory IT-22:644–654
6. Scott M (2007) Efficient implementation of cryptographic pairings. Available via DIALOG. http://ecrypt-ss07.rhul.ac.uk/Slides/Thursday/mscottsamos07.pdf
7. He D, Bu J, Zhu S, Chan S, Chen C (2011) Distributed access control with privacy support in wireless sensor networks. IEEE Trans Wirel Commun (10)10:3472–3481
8. Lin X, Sun X, Ho P, Shen X (2007) GSIS: a secure and privacy preserving protocol for vehicular communications. IEEE Trans Veh Technol 56(6):3442–3456
9. Al-Riyami S, Paterson K (2003) Certificateless public key cryptography. Advances in cryptology-ASIACRYPT 2003. Lecture notes in computer science, vol 2894. Springer, New York, pp 452–473
10. Yum D H, Lee P J (2004) Generic construction of certificateless signature. In: 9th Australasian conference of information security and privacy. Lecture notes in computer science, vol 3108. Springer, New York, pp 200–211
11. Huang X, Mu Y, Susilo W, Wong D, Wu W (2007) Certificateless signature revisited. In: Proceedings of the 12th Aaustralasian conference on information security and privacy. Lecture notes in computer science, vol 4586. Springer, Berlin, pp 308–322
12. Li X, Chen K, Sun L (2005) Certificateless signature and proxy signature schemes from bilinear pairings. Lith Math J 45:76–83
13. He D, Chen J, Zhang R (2012) An efficient and provably-secure certificateless signature scheme without bilinear pairings. Int J Commun Syst 25(11):1432–1442
14. Yap W, Heng S, Goi B (2006) An efficient certificateless signature scheme. In: International federation for information processing. LNCS, vol 4097. Springer, Berlin, pp 322–331

# Chapter 3
# Joint Physical–Application Layer Security

**Abstract** Due to the explosive growth of mobile data traffic, wireless multimedia content delivery turns to be an important application in D2D communications. In this chapter, we focus on the security for wireless multimedia transmission over D2D communications. The existing security methods consider physical-layer and application-layer security technologies independently while both of them have significant impact on security performance. Thus, we propose a joint framework involving both the physical-layer and application-layer security technologies. It is well known that the network resources are expected to be utilized efficiently by exploiting the secrecy capacity and signal processing technologies at the physical-layer as well as the authentication and watermarking strategies at the application-layer. We also design a novel network-aware security protection strategy which is implemented in a distributed manner. It achieves a tradeoff between the security level and communication overhead, which facilitates its deployment in large-scale wireless multimedia systems.

The organization of this chapter is as follows. Firstly, we give an overview of physical-layer security in wireless networks in terms of theoretical secrecy capacity, channel-based key agreement, and channel-based authentication in Sect. 3.1. Then, Sects. 3.2 and 3.3 introduce the physical-layer and application-layer security technologies for wireless multimedia delivery over D2D communications, respectively. Subsequently, a joint physical–application layer security framework is proposed in Sect. 3.4. Finally, we discuss future research directions and conclude this chapter in Sect. 3.5. Note that the majority of the contents of this chapter are based on our previous work [1].

## 3.1 Overview of Physical-Layer Security

Physical-layer security is now emerging as a new secure communication means to achieve perfect secrecy by exchanging confidential messages over a wireless medium in the presence of unauthorized eavesdroppers, without relying on higher-layer encryption [2]. Pioneered by the work of Shannon [3] and Wyner [4], the

© The Author(s) 2016

A. Zhang et al., *Security-Aware Device-to-Device Communications Underlaying Cellular Networks*, SpringerBriefs in Electrical and Computer Engineering, DOI 10.1007/978-3-319-32458-6_3

basic idea behind physical-layer security is to explore the physical characteristics of wireless channels to prevent the eavesdropper from intercepting the ongoing wireless communications. It is primarily achieved by designing transmit coding strategies intelligently without a secret key, or developing secret keys over public channels based on channel characteristics. Recently, physical-layer security has been widely studied in wireless network [1, 2, 5, 6].

In particular, [2] gives a comprehensive overview of physical-layer security in multiuser wireless networks. It describes the evolution of secure transmission strategies from point-to-point channels to multiple-antenna systems, followed by generalizations to multiuser broadcast, multiple-access, interference, and relay networks. The survey also covers secret key generation, channel coding, and message authentication based on the physical-layer mechanisms. To avoid reinvent the wheel, we refer to [2] for details. In this work, we review physical-layer security techniques in specific wireless networks, i.e., cognitive radio networks, cooperative wireless networks, and wireless sensor networks, aiming to shed lights on mechanism designs in D2D communications. Generally, the existing works fall into three categories: theoretical secrecy capacity, channel-based key agreement, and channel-based authentication.

### 3.1.1   Secrecy Capacity

Secrecy capacity is defined to measure the maximum transmission rate at which the eavesdropper is unable to recover any information [7]. It is developed from an information-theoretical prospective and denoted by the difference between the capacity of the channel from source to destination and that of the channel from source to eavesdropper. If the secrecy capacity is negative, the eavesdropper will succeed in decoding the source signal and an intercept event occurs in this case [7]. In order to measure the outage probability of the secure communications, secrecy outage probability is introduced, which is defined as the probability that the secrecy capacity is lower than the minimum secrecy requirement of the user.

Secrecy capacity and secrecy outage probability have been extensively studied in cognitive radio networks and cooperative communications [8–14]. Bassily et al. [8] and Shu et al. [9] survey the recent advances of physical-layer security in cognitive radio networks, including existing security attacks, countermeasures, cooperative jamming, and signal processing. Zou et al. [10] propose a user scheduling scheme to achieve multiuser diversity for improving the security level of cognitive transmissions with a primary QoS constraint. It also derives the diversity order of the proposed multiuser scheduling scheme through an asymptotic intercept probability analysis and proves that the full diversity is obtained by using the proposed multiuser scheduling.

Zou et al. [11] and Mo et al. [12] study relay selection and placement problems in cooperative wireless networks. They explore the physical-layer security for decode-and-forward (DF), amplify-and-forward (AF), and randomize-and-forward (RF) relay strategies, respectively. By conducting an asymptotic intercept probability analysis, they show that no matter which relaying protocol is considered (i.e., AF and DF), the traditional and proposed optimal relay selection approaches both achieve the diversity order. In [13] and [14], the authors address secure communications of one source–destination pair with the help of multiple cooperating relays in the presence of one or more eavesdroppers. They study the optimal relay weights to maximize secrecy capacity, under the power constraint for the DF and Cooperative Jamming (CJ) protocols. The authors also study the optimal weights to minimize the total power subject to a secrecy rate constraint.

The abundant works on physical-layer security in cognitive radio networks and cooperative wireless network provide rich and successful experiences for investigation of secrecy capacity in D2D communications. This is because relaying is an important function for D2D communications, especially for multimedia content delivery.

Physical-layer security issue is also addressed in the smart grid and wireless sensor networks [15–17]. Lee et al. [15] describe a random spread-spectrum-based wireless communication scheme that can achieve both fast and robust data transmission with a deep analysis of attacks and threats on the physical-layer of wireless communication in smart grid. Li et al. [16] propose novel approaches to enhance the security of wireless sensor networks by developing two randomized array transmission schemes at the physical-layer. The transmission secrecy is guaranteed by either the inherent ambiguities of MIMO blind equalization or the intentionally created ambiguities. In [17], the problem of binary hypothesis testing is considered in a bandwidth-constrained low-power wireless sensor network operating over insecure links. They consider both hard decision (binary quantization) and soft-decision cases (multilevel quantization). The proposed method may be considered as a PHY-layer security scheme for distributed detection applicable in D2D communications due to its low computational complexity and no communication overhead.

Secrecy capacity of cellular networks is investigated in [18–21]. Authors in [18, 19] study the information-theoretic secrecy performance in large-scale cellular networks based on a stochastic geometry framework. They consider information exchange between base stations, and characterize its impact on the achievable secrecy rate of an arbitrary downlink transmission with a certain portion of the mobile users acting as potential eavesdroppers. The results may help to investigate secrecy capacity of D2D communications underlaying cellular networks due to the fact that the D2D users may be considered as potential eavesdroppers. Geraci et al. [20, 21] analyze the secrecy capacity with regularized channel inversion precoding by combing tools from stochastic geometry and random matrix theory. Their analysis shows that the secrecy rate in a cellular network does not grow monotonically with the transmit power in multi-cell environments. The results may be used as guidelines for physical-layer security exploitation and power control in D2D communications. The existing works are concluded in Table 3.1.

**Table 3.1** The existing works on secrecy capacity

| Application scenarios | Literature | Techniques | Improved performances | Experiences for D2D communications |
|---|---|---|---|---|
| Cognitive radio networks | Bassily et al. [8] and Shu et al. [9] | Survey | Security attacks, countermeasures, cooperative jamming, and signal processing | The schemes provide experiences for D2D assisted communications |
|  | Zou et al. [10] | Asymptotic intercept probability analysis | Multiuser diversity |  |
| Cooperative wireless networks | Zou et al. [11] and Mo et al. [12] | Asymptotic intercept probability analysis | Diversity order |  |
|  | Dong et al. [13] and Li et al. [14] | Optimization | Maximize secrecy capacity under power constraint |  |
| Smart grid | Lee et al. [15] | Random spread-spectrum | Fast and robust data transmission | The energy and bandwidth constraint solutions may be lent to D2D communications |
| Wireless sensor networks | Li et al. [16] | Randomized array transmission | Transmission secrecy |  |
|  | Soosahabi and Naraghi-Pour [17] | Binary hypothesis testing | Low computational complexity and no communication overhead |  |
| Cellular networks | Wang et al. [18, 19] | Stochastic geometry | Location-based secrecy capacity | The results provide guidelines for D2D communications |
|  | Geraci et al. [20, 21] | Stochastic geometry and random matrix theory | Deployment density-based secrecy capacity and secrecy outage probability |  |

### 3.1.2  Channel-Based Key Agreement

Due to Maurer's seminal work in [22], a secret key agreement protocol is proposed to be implemented over a noiseless but authenticated and publicly observable two-way channel in the presence of a passive eavesdropper. Then, pioneered by the work of Azimi-Sadjadi et al. in [23], the physical-layer characteristics of wireless networks is coupled with key generation algorithms. Since then, there have been active researches on key agreement from channel measurements [24–35]. Usually, the channel-based key agreement schemes are composed by four steps: channel measurement, quantization, information reconciliation, and privacy amplification [36]. The existing approaches focus on one or several steps for the improvement of key agreement performances, as shown in Table 3.2.

The works in [24, 25] show that channel response from OFDM subcarriers can achieve higher bit generation rate for both static and mobile cases in real-world scenarios. Liu et al. [24] develop a Channel Gain Complement (CGC) assisted secret key extraction scheme to cope with channel non-reciprocity encountered in practice. Xi et al. [25] use a validation recombination mechanism to obtain high security level of the keys and fast key generation rate. There are works using phase information for key extraction in multipath channel [26, 27]. The approach in [26] is particularly attractive in wideband channels which exhibits enough statistically independent degrees of freedom (DoF) for secure key generation. The scheme in [27] enjoys a high flexibility and can be applied in both static and mobile environments due to the randomness of the phases. Different from the works which investigate a particular channel measurement, [37] derives upper and lower bounds on the secret key rate of a three-node cooperative wireless system disregarding its channel characteristics. The research seeks a tradeoff between security and protocol efficiency in the joint design of advantage distillation, information reconciliation, and privacy amplification.

Another line of researches focuses on exploring signal strength for shared key generation [28–35, 38, 39]. Croft et al. [28] and Patwari et al. [29] focuse on reducing non-reciprocities of the channel measurements for improvement of key generation rate. Croft et al. [28] introduces ranking to remove the differences in the unknown transmitter and receiver characteristics while [29] addresses the non-reciprocities problem by fractional interpolation filtering. Zeng et al. [30] propose a secret key generation protocol based on multiple-antenna. The scheme shows the tradeoff between bit generation rate and bit agreement ratio when using multilevel quantization.

Mathur et al. [31] and Zhu et al. [32] use level crossing for secret bits extraction. Mathur et al. [31] design a protocol to resist cryptanalysis of an eavesdropping adversary and a spoofing attack by an active adversary without requiring an authenticated channel, i.e., it works against passive adversary as well as active adversary. Zhu et al. [32] propose an online parameter learning mechanism to adapt to different channel conditions such that the scheme can be executed in noisy, outdoor vehicular environments. In [33] and [34], the authors exploit the difference between two RSS values to extract the secret bits. Liu et al. [33] focus on group communication while [34] emphasizes the scheme's ability of resisting active attack. Premnath et al. [35] develops an adaptive lossy quantizer to design an environment adaptive secret key generation scheme, which generates high entropy bits at a high bit rate in mobile environment.

All the above RSS-based key agreement schemes are only applicable in mobile environment. In order to improve bit rate in static wireless networks, [38] integrates opportunistic beamforming and frequency diversity for key generation. Whereas, it

**Table 3.2** The existing key agreement schemes

| Literature | Channel measurements | Techniques | Improved performances | Application scenarios | Experiences for D2D communications |
|---|---|---|---|---|---|
| Liu et al. [24] and Xi et al. [25] | CSI and OFDM | Filtering for quantization; CGC for information reconciliation | Bit rate; Security level | Mobile/static | Quantization method |
| Sayeed and Perrig [26] and Wang et al. [27] | Phase differences | Seeking optimal quantization strategy | Entropy; Security level; Energy | Mobile/static | Quantization method |
| Croft et al. [28] and Patwari et al. [29] | RSS value | Ranking or fractional interpolation filtering for channel measurement | Bit rate | Mobile | Key agreement from the RSS of D2D communications by exploiting the randomness and symmetric characteristics of D2D channels |
| Zeng et al. [30] | RSS value | Multiple-antenna and multilevel quantization | Bit rate; Bit agreement ratio | Mobile | |
| Mathur et al. [31] and Zhu et al. [32] | RSS differences | Level crossing for quantization | Bit rate; Robustness | Mobile | |
| Liu et al. [33] and Zan et al. [34] | RSS differences | RSS fading trend detection for quantization | Bit mismatch; Security level | Mobile | |
| Premnath et al. [35] | RSS value | Adaptive lossy quantization | Entropy | Mobile | |
| Azimi-Sadjadi et al. [23] | Deep fading | Deep fade detection for randomness extraction; Fuzzy information reconciliation | Robustness | Mobile | |
| Gollakota and Katabi [40] | RSS value | Repeated transmission | Bit rate; Bit agreement ratio | Mobile/static | |
| Huang and Wang [38] | RSS value | Opportunistic beamforming and frequency diversity | Entropy; Bit rate | Mobile/static | |
| Wilhelm et al. [39] | RSS value | Multilevel quantization | Robustness; Security level | Frequency-selective channel | |

requires two antennas at the transmitter for the introduction of channel fluctuations. Gollakota and Katabi [40] introduces iJam to ensure that an eavesdropper cannot even demodulate a wireless signal not intended for it. The scheme is channel-independent but it is only effective for an OFDM-based system and the secrecy depends on the statistical characteristics of transmitting data. Wilhelm et al. [39] demonstrate the applicability of a key generation protocol that takes advantage of frequency-selectivity multipath fading in static indoor WSNs. However, this approach needs to increase channel amount or channel spacing for tradeoffs between channel deviations and security level.

Due to the fact that D2D communications enable two proximity devices to setup connections directly, D2D channels have good symmetric characteristics, which may be advantageous for symmetric key agreement from the channel measurements.

### 3.1.3 Physical-Layer Authentication

A very important issue of physical-layer security is to ensure that confidential messages are decoded only by their intended receivers and to enable the receiver to detect whether it was forged or illegitimately modified by attackers. Consequently, message and entity authentication are necessary components of physical-layer security. These operations are usually performed at higher layers, i.e., network layer or application-layer, with recent interest in devising physical-layer counterparts. In recent years, a large number of studies on physical-layer authentication have appeared, including message authentication and entity authentication [41–52], as shown in Table 3.3.

Physical-layer message authentication is investigated in [41–43] from an information-theoretic analysis. Maurer [41] provides information-theoretic lower bounds on an opponent's probability of cheating in one-way message authentication by interpreting message authentication as a hypothesis testing problem. In [42], information-theoretic lower and upper bounds on the opponent's success for both single and multiple-message scenarios is studied for message authentication over noisy channels, where impersonation and substitution attacks happen. Yu et al. [43] introduce a general analysis and design framework for physical-layer authentication by superimposing a carefully designed secret modulation on the waveforms. In this way, the authentication is added to the signal without requiring additional bandwidth.

Additionally, physical-layer assist message authentication is studied in the smart meter system and distributed ad hoc wireless sensor networks [44, 45]. The authors integrate the conventional application-layer message authentication schemes and the physical-layer authentication mechanisms by taking advantage of temporal and spatial uniqueness of physical-layer channel responses, aiming to achieve fast authentication while minimizing the packet transmission overhead. Fingerprint embedding at the physical-layer for authentication is shown to deliver high probability of authentication without additional bandwidth and with minimal impact on

**Table 3.3** State-of-the-art of physical-layer authentication

| Authentication | Literatures | Methods | Application scenarios | Improved performances | Experiences for D2D communications |
|---|---|---|---|---|---|
| Message authentication | Maurer [41] | Information-theoretic analysis | Wireless networks | lower bounds on an opponent's probability of cheating | Physical-layer assist message authentication in D2D communications may be an optimal option |
| | Lai et al. [42] | Information-theoretic analysis | Wireless networks | lower and upper bounds on the opponent's message authentication | |
| | Yu et al. [43] | Secret modulation on the waveforms | Wireless networks | Without requiring additional bandwidth | |
| | Wen et al. [44, 45] | Physical-layer channel responses | Wireless sensor networks | Fast authentication | |
| | Verma et al. [53] and Yu et al. [54] | Fingerprint | Wireless networks | Secure and bandwidth-efficient | |
| Entity authentication | Shi et al. [46, 47] | Channel characteristics | Wireless body area network | Lightweight authentication | The entity authentication schemes for mobile systems and wireless networks may be applicable in D2D communications |
| | Du et al. [48] | Channel impulse response | Wireless fading channel | High successful authentication rate and low false acceptance rate | |
| | Liu et al. [49] | Two dimensional quantization | Wireless networks | High successful authentication rate | |
| | Tugnait [50] | Comparisons of random signals | Wireless networks | Low burden of symbol timing synchronization | |
| | Hou et al. [51] | Time-varying carrier frequency offset | Mobile systems | Continuous and dynamic device verification | |
| | Borle et al. [52] | Authentication tags generated by one-way hash chain | Cognitive radio networks | Reliable authentication | |

bit error rate in [53, 54]. Yu et al. [54] describe a general framework for fingerprint embedding at the physical-layer in order to provide message authentication, which is secure and bandwidth-efficient. A deliberate fingerprint embedding for message authentication is designed to enhance security by making the authentication tags less accessible to the adversaries.

In D2D communications, physical-layer assist message authentication may be an advantageous option because D2D links are usually established opportunity and the systems have the characteristics of dynamics. The D2D users may combine application-layer message authentication and physical-layer authentication schemes to achieve reliable and fast authentication.

Another line of works focuses on physical-layer entity authentication by exploring the wireless channel characteristics [46–49]. Shi et al. [46, 47] explore the heterogeneous channel characteristics among the collection of on-body channels during body motion, and between on-body and off-body communication channel to achieve lightweight body area network authentication (BANA). BANA adopts clustering analysis to differentiate the signals from an attacker and a legitimate node in [46] while MASK-BAN in [47] achieves authentication through multi-hop stable channels, which greatly reduces the false positive rate as compared to existing work. Du et al. [48] propose two physical-layer challenge-response authentication mechanisms by fully utilizing the randomness, reciprocity, and location de-correlation features of the wireless fading channel to hide/encrypt the challenge-response messages at the physical-layer. In order to reduce the false alarm rate and achieve more reliable spoofing detection for the CIR (channel impulse response)-based physical-layer authentication, [49] integrates additional multipath delay characteristics into the scheme and introduces a two dimensional quantization method to tolerate these random errors of CIRs.

Physical-layer entity authentication can also be achieved by exploiting other physical-layer characteristics [50–52]. Tugnait [50] authenticates users by comparing two random signal realizations to ascertain whether they have identical power spectral densities. Hou et al. [51] exploit the time-varying carrier frequency offset (CFO) associated with each pair of wireless communications devices to achieve physical-layer authentication. In the proposed authentication scheme, Kalman filtering is employed to predict the current CFO value by tracking the past CFO variation. The current CFO estimates is compared with the Kalman predicted CFO using hypothesis testing to determine whether the signal has followed a consistent CFO pattern, thus achieving the proposed authentication. Borle et al. [52] generate authentication tags using a one-way hash chain and embeds tags through constellation shift for cognitive networks. Additionally, an optimal strategy for tag embedding is provided which minimizes the tag bit detection probability for a given tag power.

## 3.2 Physical-Layer Security for Wireless Multimedia Delivery

In wireless multimedia communication system, secrecy capacity and information processing approach are two important technologies to realize physical-layer security. Thus, we briefly introduce the two methods in this section.

### 3.2.1 Security Capacity

As stated in Sect. 3.1, the secrecy capacity (SC) has gained tremendous attention in recent years. Generally, SC provides a theoretical maximum transmission rate for any multimedia content delivery.

In the case that the authorized transmitter can obtain the channel states of the authorized receiver and unauthorized user, SC is defined as the upper bound of the channel capacity when the authorized receiver only accesses the CSI of the authorized transmitter [55]. However, in some practical application scenarios, it is difficult to obtain the full CSI. Thereafter, [56] investigates SC in the presence of imperfect CSI. Specifically, two legal users communicate over a quasi-static fading channel and an unauthorized user accesses their transmissions via another quasi-static fading channel. The authors provide a close-expression for the upper bound of SC in this case by characterizing the role of fading channel.

It is observed that SC is an average-information measurement for both the full and imperfect CSI cases, and its upper bound can be designed by utilizing different kinds of technologies in physical-layer.

### 3.2.2 Information Processing Approach

Information processing approach (IPR) provides various technologies to achieve SC [55]. The key technology of IPR is to distinguish the channel feedback information from the authorized to unauthorized users at the initial step of communications. Generally, noises are injected to the authorized users to prevent the unauthorized users from estimating the channel state. To this end, there are two widely methods to realize scalable try-test channel estimation:

- *Multiple antennas*. Each authorized user is equipped with multiple-antenna thus the authorized transmitter can obtain its channel state information from the authorized receiver's feedback.
- *Advanced coding*. Advanced coding is a general term compared to traditional method. We take the spread-spectrum coding (SSC) as an example. In SSC, a signal is spread by a pseudo-noise sequence over a wide frequency band.

The main characteristic is that its key size is small because the key space is compressed by different key sequences in SSC, while conventional cryptographic methods require a large key space.

Due to the artificial noises, the quality of the channel estimation obtained by the unauthorized users is reduced while that of the authorized users can be retained since the scalable try-best estimation in the above two methods.

## 3.3 Application-Layer Security for Wireless Multimedia Delivery

The application-layer security is usually guaranteed by authentication and watermarking for the wireless multimedia communications. We give a brief introduction and some classical applications of these methods in this section.

### 3.3.1 Authentication

The aim of authentication is to determine whether the received content (1) has been sent by the legal transmitter, and (2) has not been altered during the transmission process [57]. In other words, multimedia authentication addresses two concerns: (1) who transmits the multimedia data, and (2) whether the received multimedia data has been changed or not.

The existing end-to-end media authentication schemes can be classified into *stream-based* and *content-based* techniques [58]. The *stream-based authentication* directly authenticates at the multimedia stream while the *content-based authentication* authenticates the media at the content level. Stream-based authentication can be further divided into error correction coding (ECC) based methods and graph-based methods [59]. As the basic data processing unit is block composed by a number of packets, the digital signature is coded with an erasure code and then dispersed over all the packets of the block [60]. All received packets are able to be authenticated if and only if the amount of lost packets is lower than a specific threshold defined by the authentication level. However, this method suffers from high computational overhead because of the erasure coding. Additionally, the receiver delay may be high due to the minimum packets required for authentication.

Li et al. [59] introduce the basic framework for graph-based authentication. In this method, authentication is realized after receiving the last packet, and each packet is appended with its hash. Meanwhile, the signature is computed over all of the hashes of the packets. Sun et al. [58] and Gennaro and Rohatgi [61] show that a simple hash chain is suffice for authentication. This method has the advantages of relatively low overhead and low receiver delay while suffering from high transmission delay and packet loss sensitive. In order to solve this problem,

[62, 63] propose a butterfly graph for stream authentication, which is robust against both random and burst packet loss. The scheme outperforms the existing works in terms of overhead, verification probability, and delay. Readers are referred to [59, 64, 65] for more works on graph-based stream authentication.

*Content-based authentication* aims to authenticate the media at the content level. Sun et al. [58] demonstrate that authenticating media content can be achieved by authenticating the invariant features of the content. Usually, digital signature generation or signing is employed to achieve multimedia content authentication, which has to tolerate a certain false acceptance ratio (FAR) and false rejection ratio (FRR) [66, 67]. Consequently, due to its limitations on the acceptable FAR and FRR, the content-based authentication scheme's usefulness is restricted in many applications.

### 3.3.2   Watermarking

Watermarking is a kind of signature embedded in the multimedia object by the owner, which helps to realize the copyright protection, content authentication, tamper detection, etc. Usually, the watermarking depends on the private key of the detector, and is tested and examined by the decoder. There have been substantial watermarking systems from various perspectives, and readers are referred to [60, 68] for a comprehensive survey and tutorial. Generally, three cases should be considered when embedding the watermarking in a multimedia content.

- *Pre-encoding*: Watermarking is implemented before the multimedia content encoding in this case. As encoding is considered as an authorized operation, the embedded watermark may be destroyed or removed if the encoded stream employs a large quantization. Therefore, quantization plays a very important role in this case.
- *Inter-encoding*: In this case, watermarking is involved in the encoder and the encoding function is modified accordingly. The quality of the watermark is not destroyed by the encoder and its implementation is simple but it suffers from high complexity of the decoder.
- *Post-encoding*: Watermarking is implemented after the encoding by scheduling, transcoding, or resource allocation, etc. Due to the fact the watermarking is sensitive to the environment dynamics, this method should be carefully employed since it takes up too large encoded rate, which is fatal for wireless networks.

In summary, watermarking helps to protect the multimedia content while bringing some non-ignorable computational complexity burden for the encoder or the decoder. Consequently, a watermarking scheme should consider the tradeoff between the computational complexity, signal robustness, and information reliability during the signal processing [69].

## 3.4  Joint Physical–Application Layer Security Scheme

In this section, we describe the joint physical–application layer security scheme from the aspects of framework description, security-aware packetization, and a specific joint scheme.

### 3.4.1  Framework Description

Generally, the total framework of the proposed scheme is based on an error-resilient encoder and decoder, as shown in Fig. 3.1 [1]. In specific, the multimedia data is first passed to the security capacity part at the transmitter. This part estimates and analyzes the maximal channel capacity and outputs the maximum transmission rate including source coding rate, channel coding rate, authentication rate, and watermarking rate. Resource allocation or scheduling is also implemented in this part. The encoding part plays the role of source coding by encoding the multimedia contents according to the received input rate and the required output rate. The channel coding part constructs authentication and watermarking, which establishes, signs, and protects the media streaming before transmission.

At the receiver, the decoder detects the transmission errors and utilizes error-concealment to reduce the cost of code error. Specifically, the error detection information is sent to the packet verification part, where the false alarms may be triggered by the multimedia content errors. Usually, the threshold of the false alarm is determined by the authentication request.

The packet verification employs traditional hash functions and is implemented via packet-by-packet operation. The verification results are processed at the decoder,

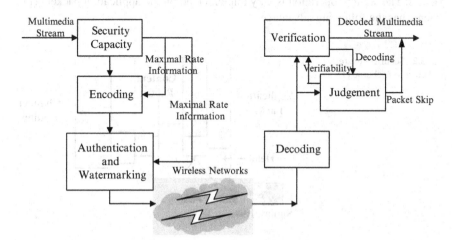

**Fig. 3.1**  Joint physical–application layer security framework [1]

and the nonverifiable packets are skipped. The practical wireless network architecture consists of several layers, thus it is beneficial to interaction between the encoder and decoder while inevitably leading to data redundancy. Thereafter, it is required to achieve the optimal tradeoff between data redundancy and security level by security-aware cross-layer designing.

### 3.4.2  Security-Aware Packetization

The stream is packetized according to their content priority and difference in order to provide the dynamic packet protection for the encoded stream [58]. Conceptually, the packetization scheme should demonstrate the differences of the importance and category of the packet based on the multimedia content. Two kinds of packets are introduced in [58]:

1. Content packet. Compressed multimedia stream is correct if and only if all the received packets are corrected.
2. Application packet. The application packet is grouped into different packet blocks for a or multiple applications.

The existing packetization methods re-shuffle the errors across all the packets in a packet block to reduce the decode errors, where block is a basic decode unit for error detection and correction. As each block contains many content packets and application packets, it is difficult to identify the importance and priority of the packet in the block, which causes the inefficiency of the error detection.

In order to address the above issues, a new security-aware packetization scheme is proposed, motivated by the definition of smart error detection in [59]. As shown in Fig. 3.2 [1], each block contains only one application packet and multiple content packets. The signing operation is only implemented on the application packet. This scheme has the following advantages:

**Fig. 3.2** Security-aware packetization scheme [1]

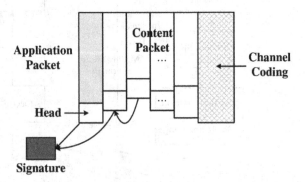

- The importance and priority of the packets are obtained as the operations of authentication and the watermarking directly relate to the multimedia content since the application packet only depends on the multimedia content;
- As only one application packet is processed in each block, the computational complexity of the application-layer security is reduced dramatically;
- The security capacity information can fully and flexibly approach the limit;
- As the scheme is distributed to several content packets instead of only one, the system is more robust to the code error.

### 3.4.3 Joint Scheme

The joint physical–application layer security scheme is shown in Fig. 3.1. The decoder feedbacks the error report for each block to the encoder, which may suffer from the unexpected attack. In order to address this issue, the traditional cryptography approach may be employed to guarantee transmission security and duplicate transmission fashion is adopted to ensure that this report can be correctly received at the transmitter.

When the errors have been detected at the transmitter, it first determines whether the error value is above the threshold or not. If the error is lower than the given threshold, the packets in this block will be skipped for verification. Otherwise, authentication and watermarking are reconstructed by reducing the strength of security level. This scheme achieves security at the cost of substantial communication overhead.

As a result, we propose a scheme which considers the physical- and application-layer security jointly. Firstly, the secrecy capacity, denoted by $C$, is estimated before the transmission. Intuitively, the estimation frequency is different for different network environments. Generally, the update frequency for the dynamic wireless networks is higher than that of the static wireless networks. Note that information update does not increase the communication overhead because it is conducted at the transmitter. Subsequently, in order to improve the content robustness, sophisticated channel coding technologies are introduced in the scheme, and the corresponding rate is denoted by $C_c$. The above two steps are implemented at the physical-layer. Then the authentication and watermarking are operated at the application-layer. The authentication rate is $C_a$ and the watermarking rate is $C_w$, and the sum of the $C_a$ and $C_w$ is not larger than $(C - C_c)$. In the circumstances that authentication and watermarking have the same importance, the maximal authentication or watermarking rate is $(C - C_c)/2$. The users can choose the weight arbitrarily according to their multimedia application scenarios. The readers are referred to [62] for more descriptions on the impact of the allocated rate on authentication.

Numerical simulation results have demonstrated the efficiency of the proposed joint physical–application layer security strategy. The readers are referred to [9].

## 3.5   Conclusions and Outlook

In this chapter, we propose a framework jointly considering both the physical-layer and application-layer security technologies. By exploiting the secrecy capacity and signal processing strategies at the physical-layer, and the authentication and watermarking technologies at the application-layer, the network resource is efficiently utilized. Additionally, the scalable multimedia security service can be achieved without alterations of the current network architectures. In fact, this scheme can be viewed as an example of a general cross-layer security architecture. Cross-layer security may become the new direction of the security service, and the design points may include:

- *Device authentication.* With the advances of D2D communications, entity authentication and content authentication may be jointly implemented because different devices may have different contents and security levels.
- *Content confidentiality.* The ongoing information (i.e., multimedia contents, feedback message, etc.) should be confidential to protect the privacy for both the sender and the receiver.
- *Green security.* The cross-layer security service should be operated at low computational cost and communication cost because the multimedia content is sensitive to the transmission delay.

## References

1. Zhou L, Wu D, Zheng B, Guizani M (2014) Joint physical-application layer security for wireless multimedia delivery. IEEE Commun Mag 52(3):66–72
2. Mukherjee A, Fakoorian S, Huang J, Swindlehurst A (2014) Principles of physical layer security in multiuser wireless networks: a survey. IEEE Commun Surv Tutorils 16(3):1550–1573
3. Shannon CE (1949) Communication theory of secrecy systems. Bell Syst. Tech J 28:656–715
4. Wyner AD (1975) The wire-tap channel. Bell Syst Tech J 54(8):1355–1387
5. Yuksel M, Erkip E (2007) Secure communication with a relay helping the wiretapper. IEEE Information Theory Workshop, Lake Tahoe
6. Mathur S, Reznik A, Ye C, Mukherjee R, Rahman A, Shah Y, Trappe W, Mandayam N (2010) Exploiting the physical layer for enhanced security. IEEE Wirel Commun 17(5):63–70
7. Leung-Yan-Cheong SK, Hellman ME (1978) The Gaussian wiretap channel. IEEE Trans Inf Theory 24(7):451–456
8. Bassily R, Ekrem E, He X, Tekin E, Xie J, Bloch MR, Ulukus S, Yener A (2013) Cooperative security at the physical layer. IEEE Signal Process Mag 30(5):16–28
9. Shu Z, Qian Y, Ci S (2013) On physical layer security for cognitive radio networks. IEEE Netw 27(3):28–33
10. Zou Y, Wang X, Shen W (2013) Physical-layer security with multiuser scheduling in cognitive radio networks. IEEE Trans Commun 61(12):2103–5113
11. Zou Y, Wang X, Shen W (2013) Optimal relay selection for physical-layer security in cooperative wireless networks. IEEE J Sel Areas Commun 31(10):2099–2111
12. Mo J, Tao M, Liu Y (2012) Relay placement for physical layer security: a secure connection perspective. IEEE Commun Lett 16(6):878–881

13. Dong L, Han Z, Petropulu AP, Poor HV (2010) Improving wireless physical layer security via cooperating relays. IEEE Trans Signal Process (58)3:1875–1888
14. Li J, Petropulu AP, Weber S (2011) On cooperative relaying schemes for wireless physical layer security. IEEE Trans Signal Process 59(10):4985–4997
15. Lee E, Gerla M, Oh SY (2012) Physical layer security in wireless smart grid. IEEE Commun Mag 50(3):46–52
16. Li X, Chen M, Ratazzi EP (2005) Array-transmission based physical-layer security techniques for wireless sensor networks. Proceedings of the IEEE International Conference on Mechatronics and Automation, Niagara Falls, Canada
17. Soosahabi R, Naraghi-Pour M (2012) Scalable PHY-layer security for distributed detection in wireless sensor networks. IEEE Trans Inf Forensics Secur 7(4):1118–1126
18. Wang H, Zhou X, Reed MC (2013) Physical layer security in cellular networks: a stochastic geometry approach. IEEE Trans Wirel Commun 12(6):2776–2787
Dhillon
19. Wang H, Zhou X, Reed MC (2013) On the physical layer security in large scale cellular networks. IEEE Wireless Communications and Networking Conference (WCNC)
20. Geraci G, Dhillon HS, Andrews JG, Yuan J, Collings IB (2014) A new model for physical layer security in cellular networks. IEEE International Conference on Communications, Sydney, Australia
21. Geraci G, Dhillon HS, Andrews JG, Yuan J, Collings IB (2014) Physical layer security in downlink multi-antenna cellular networks. IEEE Trans Commun 62(6):2006–2021
22. Maurer U (1993) Secret key agreement by public discussion from common information. IEEE Trans Inf Theory 39(3):733–742
23. Azimi-Sadjadi B, Kiayias A, Mercado A, Yener B (2007) Robust key generation from signal envelopes in wireless networks. ACM Conference on Computer and Communications Security (CCS), Alexandria, Virginia, USA
24. Liu H, Wang Y, Yang J, Chen Y (2013) Fast and practical secret key extraction by exploiting channel response. IEEE International Conference on Computer Communications (INFOCOM), Turin, Italy
25. Xi W, Li X, Qian C, Han J, Tang S, Zhao J, Zhao K (2014) KEEP: fast secret key extraction protocol for D2D communications. IEEE International Symposium of Quality of Service (IWQoS), Hong Kong, China
26. Sayeed A, Perrig A (2008) Secure wireless communications: secret keys through multipath. IEEE International Conference on Acoustics, Speech and Signal Processing (ICASSP), Las Vegas, NV, USA
27. Wang Q, Su H, Ren K, Kim K (2011) Fast and scalable secret key generation exploiting channel phase randomness in wireless networks. IEEE International Conference on Computer Communications (INFOCOM), Shanghai, China
28. Croft J, Patwari N, Kasera S (2010) Robust uncorrelated bit extraction methodologies for wireless sensors. ACM/IEEE International Conference on Network Protocols (ICNP), Kyoto, Japan
29. Patwari N, Croft J, Jana S,Kasera S (2010) High-rate uncorrelated bit extraction for shared secret key generation from channel measurements. IEEE Trans Mobile Comput 9(1):17–30
30. Zeng K, Wu D, Chan A, Mohapatra P (2010) Exploiting multiple-antenna diversity for shared secret key generation in wireless networks. IEEE International Conference on Computer Communications (INFOCOM), San Diego, CA, USA
31. Mathur S, Trappe W, Mandayam N, Ye C, Reznik A (2008) Radiotelepathy: extracting a secret key from an unauthenticated wireless channel. ACM International Conference on Mobile Computing and Networking (MobiCom), San Francisco, CA, USA
32. Zhu X, Xu F, Novak E, Tan CC, Li Q, Chen G (2013) Extracting secret key from wireless link dynamics in vehicular environments. IEEE International Conference on Computer Communications (INFOCOM), Turin, Italy

33. Liu H, Yang J, Wang Y, Chen Y (2012) Collaborative secret key extraction leveraging received signal strength in mobile wireless networks. IEEE International Conference on Computer Communications (INFOCOM), Orlando, FL, USA
34. Zan B, Gruteser M, Hu F (2012) Improving robustness of key extraction from wireless channels with differential techniques. International Conference on Computing, Networking and Communications, Wireless Ad Hoc and Sensor Networks Symposium (WiMob), Barcelona, Spain
35. Premnath SN, Jana S, Croft J, Gowda PL, Clark M, Kasera SK, Patwari N, Krishnamurthy SV (2013) Secret key extraction from wireless signal strength in real environments. IEEE Trans Mobile Comput 12(5):917–930
36. Ren K, Su H, Wang Q (2011) Secret key generation exploiting channel characteristics in wireless communications. IEEE Wirel Commun 18(8):6–12
37. Wang N, Zhang N, Aaron Gulliver T ( 2014) Cooperative key agreement for wireless networking: key rates and practical protocol design. IEEE Trans Inf Forensics Secur 9(2):272–284
38. Huang P, Wang X (2013) Fast secret key generation in static wireless networks: a virtual channel approach. IEEE International Conference on Computer Communications (INFOCOM), Turin, Italy
39. Wilhelm M, Martinovic I, Schmitt JB (2013) Secure key generation in sensor networks based on frequency-selective channels. IEEE J Sel Areas Commun 31(9):1779–1790
40. Gollakota S, Katabi D (2011) Physical layer wireless security made fast and channel independent. IEEE International Conference on Computer Communications (INFOCOM), Shanghai, China
41. Maurer UM (2000) Authentication theory and hypothesis testing. IEEE Trans Inf Theory 46(4):1350–1356
42. Lai L, El Gamal H, Poor HV (2009) Authentication over noisy channels. IEEE Trans Inf Theory (55)2:906–916
43. Yu PL, Baras J, Sadler B (2008) Physical-layer authentication. IEEE Trans Inf Forensics Secur 3(1):38–51
44. Wen H, Wang Y, Zhu X, Li J, Zhou L (2013) Physical layer assist authentication technique for smart meter system. IET Commun 7(3):189–197
45. Wen H, Ho PH, Qi C, Gong G (2010) Physical layer assisted authentication for distributed ad hoc wireless sensor networks. IET Inf Secur 4(4):390–396
46. Shi L, Li M, Yu S, Yuan J (2013) BANA: body area network authentication exploiting channel characteristics. IEEE J on Sel Areas Commun 31(9):1803–1816
47. Shi L, Yuan J, Yu S, Li M (2015) MASK-BAN: movement-aided authenticated secret key extraction utilizing channel characteristics in body area networks. IEEE Internet Things J 2(1):52–62
48. Du X, Shan D, Zeng K, Huie L (2014) Physical layer challenge-response authentication in wireless networks with relay. IEEE Conference on Computer Communications
49. Liu FJ, Wang X, Primak SL ( 2013) A two dimensional quantization algorithm for CIR-based physical layer authentication. IEEE International Conference on Communications
50. Tugnait JK (2013) Wireless user authentication via comparison of power spectral densities. IEEE J Sel Areas Commun 31(9):1791–1802
51. Hou W, Wang X, Chouinard J, Refaey A (2014) Physical layer authentication for mobile systems with time-varying carrier frequency offsets. IEEE Trans Commun 62(5):1658–1667
52. Borle KM, Chen B, Du W (2015) Physical layer spectrum usage authentication in cognitive radio: analysis and implementation. IEEE Trans Inf Theory Forensics Secur 10(10):2225–2235
53. Verma G, Yu P, Sadler BM (2015) Physical layer authentication via fingerprint embedding using software-defined radios. IEEE Access 3(2):81–88
54. Yu PL, Verma G, Sadler BM (2015) Wireless physical layer authentication via fingerprint embedding. IEEE Commun Mag 53(6):48–53
55. Shiu YS, Chang SY,Wu HC,Huang SCH, Chen HH (2011) Physical layer security in wireless networks: a tutorial. IEEE Wirel Commun 18(2):66–74

56. Shu Z, Qian Y, Ci S (2013) On physical layer security for cognitive radio networks. IEEE Netw 27(3):28–33
57. Simmons GJ (1992) Contemporary cryptography. IEEE, New York
58. Sun Q, Apostolopoulos J, Chen CW, Chang SF (2008) Quality-optimized and secure end-to-end authentication for media delivery. Proc IEEE 96(1):97–111
59. Li Z, Sun Q, Lian Y, Chen CW (2007) Joint source-channel-authentication resource allocation and unequal authenticity protection for multimedia over wireless networks. IEEE Trans Multimedia 9(4):837–850
60. Lai CF, Huang YM, Chao HC (2010) DLNA-based multimedia sharing system over OSGI framework with extension to P2P network. IEEE Syst J 4(2):262–270
61. Gennaro R, Rohatgi P (1997) How to sign digital streams. Proceedings of Advances in Cryptology (CRYPTO'97), pp 180–197
62. Zhou L, Chao HC, Vasilakos A (2011) Joint forensics-scheduling strategy for delay-sensitive multimedia applications over heterogeneous networks. IEEE J Sel Areas Commun 29(7):1358–1367
63. Zhang Z, Sun Q, Wong WC (2005) A proposal of butterfly-graph based stream authentication over lossy networks. Proceedings of IEEE International Conference on Multimedia and Expo (ICME)
64. Chan CF (2003) A graph-theoretical analysis of multicast authentication. Proceedings of the 23rd IEEE International Conference on Distributed Computing Systems (ICDCS'03): 155–160.
65. Lysyanskaya RT, Triandopoulos N (2004) Multicast authentication in fully adversarial networks. Proceedings of the 2004 IEEE Symposium on Security and Privacy, pp 241–248
66. Sun Q, He D, Tian Q (2006) A secure and robust authentication scheme for video transcoding. IEEE Trans Circuits Syst Video Technol 16(10):1232–1244
67. Sun Q, Ye S, Lin CY, Chang SF (2005) A crypto signature scheme for image authentication over wireless channel. Int J Image Graph 5(1):1–14
68. Vidyasagar MP, Song H, Elizabeth C (2005) A survey of digital image watermarking techniques. Proceedings of the 2005 3rd IEEE International Conference on Industrial Informatics (INDIN)
69. Wu D, Cai Y, Zhou L, Wang J (2012) A cooperative communication scheme based on dynamic coalition formation game in clustered wireless sensor networks. IEEE Trans Wirel Commun 11(3):1190–1200

# Chapter 4
# Cooperation Stimulation

**Abstract** In D2D communications the users work both as servers and clients, thus cooperation stimulation is an important technology as the system availability mainly depends on the cooperation degree of the users. Considering the particular features of D2D communication, resource-exchange-based incentive mechanism is a superior choice as it neither relies on the use of tamper-proof hardware nor monitors other's behavior as implemented in CB or RB. We propose a resource-exchange-based cooperation stimulation scheme in this chapter. Specifically, the users obtain their demanding contents at the cost of sending contents to their counterparts. By this method, contents are disseminated cooperatively with D2D communications in the network. Furthermore, as multimedia dominates the contents with Quality of Experience (QoE) as a key measurement, the cooperation stimulation mechanism is constructed for maximizing user QoE characterized by Mean Opinion Score (MOS). In the proposed scheme, the users compute their transmitter MOS and receiver MOS, and send them to the content provider (CP). Then, the CP formulates a weighted directed graph based on the network topology and connection MOS. By factorizing the graph, the content dissemination scheme is designed according to the 1-Factor with the maximum weight. Additionally, in order to realize cheat-proof, a debt mechanism is introduced in the scheme.

This chapter is organized as follows: Sect. 4.1 describes the network architecture and QoE models of the proposed scheme. In Sect. 4.2, we present the graph models including basis of graph theory, candidate graph model, and feasible graph model. A QoE-driven cooperative multimedia content dissemination scheme in D2D system is proposed in Sect. 4.3, followed by the performance analysis in Sect. 4.4. Extensive simulation results are provided in Sects. 4.5 and 4.6 concludes the chapter. Note that the majority of the contents of this chapter are based on our previous work [1].

© The Author(s) 2016
A. Zhang et al., *Security-Aware Device-to-Device Communications Underlaying Cellular Networks*, SpringerBriefs in Electrical and Computer Engineering, DOI 10.1007/978-3-319-32458-6_4

## 4.1  System Model

### 4.1.1  Network Architecture

In this chapter, we consider a D2D content dissemination system with multimedia content providers, as shown in Fig. 4.1 [1]. Consider $n$ users $\mathbb{U} = \{1, 2, \ldots, n\}$ served by the same eNB. The media content is classified into $m$ classes and we denote $\mathbb{C} = \{c_1, c_2, \ldots, c_m\}$ as the classification (i.e., movie, music, and news) set of the media content. Vector $\mathbf{X}^i = [x_1^i, x_2^i, \ldots, x_p^i]$ denotes the media possession states of the user $i$, where

$$x_k^i = \begin{cases} 1, & \text{if user } i \text{ possesses content } c_k, \\ 0, & \text{otherwise} \end{cases} \tag{4.1}$$

Let vector $\mathbf{X}_j^i \triangleq \mathbf{X}^i - \mathbf{X}^j = [x_1, x_2, \ldots, x_m], x_k \in \{0, -1, 1\}$ denote the differences of possession status between node $i$ and node $j$. Here, $x_k = 0$ denotes that they have the same possession status for content $c_k$, i.e., they both possess or both lack content $c_k$, $x_k = 1$ denotes that node $i$ possess $c_k$ while $j$ doesn't and $x_k = -1$ denotes the contrary circumstance.

### 4.1.2  QoE Model

Without loss of generality, we consider a connection $e_{ij}^k$, where node $i$ sends content $c_k$ to node $j$, as shown in Fig. 4.2. As analyzed in [2], in the connection $e_{ij}^k$, the MOS of the transmitter $i$ and the receiver $j$ are affected by content classification $c_k$, data rate $r_i^j$, and bit error rate $b_i^j$, where $r_i^j$ and $b_i^j$ denote the data rate and bit error rate from node $i$ to node $j$. Accordingly, the MOS for $j$, named receiver MOS, is estimated by

$$RQ_i^j(c_k) = \phi^j(r_i^j, b_i^j, c_k), \tag{4.2}$$

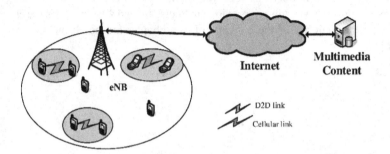

**Fig. 4.1** Network architecture for multimedia content dissemination in D2D communication

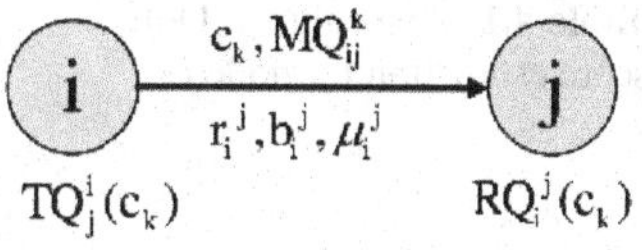

Fig. 4.2 An illustration of content exchange

and the MOS for $i$, named transmitter MOS, is estimated by

$$TQ_j^i(c_k) = \varphi^i(r_i^j, b_i^j, c_k), \tag{4.3}$$

where $\phi^j$ and $\varphi^i$ are mapping functions from the metrics to MOS for $j$ working as the receiver and $i$ as the transmitter, respectively.

The bandwidth of each sub-channel is $BHz$ and the signal-to-noise ratio (SNR) at device $j$ in connection $e_{ij}^k$ is $\mu_i^j$.[1] Then the data rate at node $i$ is computed by Host-Madsen and Zhang [3]

$$r_i^j = B \lg(1 + \mu_i^j). \tag{4.4}$$

In LTE system, M-QAM modulation signal is adopted for the symbols and we suppose uncoded $2^q$-QAM is used in the system. Accordingly, the BER at $j$ in connection $e_{ij}^k$ is [4]

$$b_i^j = 0.2 \exp\left\{-\frac{1.6 \times \mu_i^j}{2^q - 1}\right\}. \tag{4.5}$$

In Eqs. (4.4) and (4.5), both data rate and BER are the functions of SNR. Therefore, Eqs. (4.2) and (4.3) are rewritten as

$$RQ_i^j(c_k) = \phi^j(\mu_i^j, c_k) \tag{4.6}$$

and

$$TQ_j^i(c_k) = \varphi^i(\mu_i^j, c_k), \tag{4.7}$$

In order to take both the transmitter MOS and receiver MOS into consideration, we introduce connection MOS.

**Definition 4.1.** Connection MOS. Connection MOS is defined as the average MOS for the transmitter and the receiver, i.e., for a connection $e_{ij}^k$,

$$MQ_{ij}^k = \frac{TQ_j^i(c_k) + RQ_i^j(c_k)}{2},$$

where $MQ_{ij}^k$ denotes the connection MOS of $e_{ij}^k$.

---

[1] $\mu_i^j$ is estimated by node $j$ through channel state information estimation with fixed transmission power and noise variance [3].

**Table 4.1** Algorithm 1: QoE score estimation for node $i$

| | |
|---|---|
| 01: | **Input:** |
| 02: | Neighbor set $\mathbb{N}^i, \mu_j^i, \mathbf{X}_j^i$, for all $j \in \mathbb{N}^i$ |
| 03: | **Output:** |
| 04: | $\mathbb{R}^i$ and $\mathbb{T}^i$; |
| 05: | **Procedures** |
| 06: | $\mathbb{R}^i = \varnothing, \mathbb{T}^i = \varnothing$; |
| 07: | **For all** $j \in \mathbb{N}^i$ **do** |
| 08: | **for**$(k = 1 : t)$ |
| 09: | **if** $(\mathbf{X}_j^i(k) = 1)$ |
| 10: | $TQ_j^i(c_k) = \varphi^i(\mu_j^i, c_k)$, |
| 11: | $\mathbb{T}^i = \mathbb{T}^i \cup \{TQ_j^i(c_k)\}$; |
| 12: | **else** |
| 13: | **if** $(\mathbf{X}_j^i(k) = -1)$ |
| 14: | $RQ_j^i(c_k) = \phi^i(\mu_j^i, c_k)$, |
| 15: | $\mathbb{R}^i = \mathbb{R}^i \cup \{RQ_j^i(c_k)\}$. |
| 16: | **endif** |
| 17: | **endif** |
| 18: | **endfor** |
| 19: | **endfor** |

In order to stimulate cooperation, the user that receives content from its neighbor is required to transmit content to others for exchange, i.e., each user works both as the server and customer. For each user $i \in \mathbb{U}$, it keeps two score sets $\mathbb{T}^i$ and $\mathbb{R}^i$, which denote transmitter MOS and receiver MOS, respectively. Let $\mathbb{N}^i = \{i_1, i_2, \ldots, i_l\}, i_k \in \mathbb{U}$ denote node $i$'s neighbor set. Then $\mathbb{T}^i$ and $\mathbb{R}^i$ are estimated by **Algorithm 1**, as shown in Table 4.1. The notations throughout this chapter are summarized in Table 4.2.

## 4.2   Graph Models

We firstly describe some basis of graph theory in this section for the sake of completeness. Then we define candidate graph model and feasible graph model for the D2D content dissemination system.

### 4.2.1   Basis of Graph Theory

We only review some basis of graph theory in this section. For more comprehensive introduction, please refer to [5].

**Table 4.2** Notations and description

| Notation | Description | Notation | Description |
|---|---|---|---|
| $e_{ij}^k$ | The connection that $i$ sends content $c_k$ to node $j$ | $\mathbf{X}_j^i$ | The possession differences between node $i$ and node $j$ |
| $RQ_i^j(c_k)$ | The MOS for $j$ performing as the receiver in connection $e_{ij}^k$ | $\mathbb{R}^i$ | Receiver MOS set for node $i$ |
| $TQ_j^i(c_k)$ | The MOS for $i$ performing as the transmitter in connection $e_{ij}^k$ | $\mathbb{T}^i$ | Transmitter MOS set for node $j$ |
| $MQ_{ij}^k$ | Connection MOS for $e_{ij}^k$ | $\mathbb{G}_\tau$ | $\tau$-feasible graph |
| $r_i^j$ | The data rate achieved by node $i$ in connection $e_{ij}^k$ | $w(\mathbb{G})$ | The weight for graph $\mathbb{G}$ |
| $b_i^j$ | Bit error rate ($b_i^j$) achieved by node $i$ in connection $e_{ij}^k$ | $w_{ij}^k$ | The weight for connection $e_{ij}^k$ |
| $\mu_i^j$ | The signal-to-noise ratio at receiver $j$ in connection $e_{ij}^k$ | $p(\mathbf{A})$ | The permanence of matrix $\mathbf{A}$ |
| $\mathbb{N}^i$ | Node $i$'s neighbor set | $\mathbb{P}(\mathbf{A})$ | The pseudo-permanence of matrix $\mathbf{A}$ |

A graph $\mathbb{G} = \{V, E\}$ is composed by vertex set V and edge set E. The degree of a vertex is the amount of edges connected with the vertex. If all the edges E are directed, the graph is a directed graph. The in-degree and out-degree of a vertex in directed graph $\mathbb{G}$ refers to the amount of edges point to and point out from the vertex, respectively. A directed graph $\mathbb{H}$ is a subgraph of the directed graph $\mathbb{G}$ if the vertices and edges of $\mathbb{H}$ are vertices and edges of $\mathbb{G}$, respectively, i.e., $V(\mathbb{H}) \subseteq V(\mathbb{G})$ and $E(\mathbb{H}) \subseteq E(\mathbb{G})$. In addition, if each vertex of $\mathbb{G}$ is a vertex of $\mathbb{H}$, then $\mathbb{H}$ is a *spanning subgraph* of $\mathbb{G}$. A directed graph is regular if all the vertices have the same in-degree and out-degree. *1-Factor* of a directed graph $\mathbb{G}$ is defined as follows:

**Definition 4.2. 1-Factors**. The 1-Factor of a directed graph $\mathbb{G}$ is defined as its spanning subgraph which is regular of degree 1. All the 1-Factors of the graph are defined as 1-Factors.

A graph with $n$ vertices can be denoted by its adjacency matrix. The adjacency matrix of $\mathbb{G}$ is an $n \times n$ matrix $\mathbf{A} = [a_{ij}]$, where

$$a_{ij} = \begin{cases} 1, & \text{if } (i,j) \in E \\ 0, & \text{otherwise} \end{cases}.$$

The *permanence* of a matrix $\mathbf{A}$ is defined as

$$p(\mathbf{A}) = \sum_{(j)} a_{1j_1} a_{2j_2} \dots a_{nj_n},$$

where the summation is taken over all the $n!$ permutations $(j) = (j_1, j_2, \dots j_n)$. We define the non-zero item of the summation as a *component* of the permanence.

By exploring the characteristics of permanence, *Lemma 4.1* provides a method for seeking the 1-Factors of a graph [5, 6].

**Lemma 4.1 (Zhang et al. [1]).** *The amount of 1-Factors of a directed graph* $\mathbb{G}$ *equals to the permanence of its adjacency matrix* **A**. *The* **component** *of the permanence and the 1-Factor formulate one-to-one map.*

*Proof.* Note that a permutation $a_{1j_1} a_{2j_2} \ldots a_{nj_n} = 1$ corresponds to a spanning subgraph of $\mathbb{G}$, denoted by $H_1$. It can be found that $H_1$ is composed by edge $(1, j_1), (2, j_2), \ldots (n, j_n)$, each element (vertex) appears twice in all the directed sequence, once in the first element and once in the second one. As a result, the in-degree and the out-degree for each vertex of $H_1$ is one. Therefore, $H_1$ is a spanning subgraph of $\mathbb{G}$ with regular of degree 1, i.e., 1-Factor of $\mathbb{G}$. On the other hand, for each 1-Factor of $\mathbb{G}$, there is a corresponding non-zero component and only this component in the permanence of **A**. Consequently, the non-zero component and 1-Factor formulate one-to-one map.                                                    ∎

A graph has no 1-Factor if the permanence of its adjacency matrix is zero.

### 4.2.2  Candidate Graph Model

In the D2D content dissemination system, as different devices possess and request the same or/and different contents, not all the device within communication radius can establish a link. We introduce candidate graph in the system to take all these possible connections into considerations. Firstly, we introduce the concepts of candidate peers and candidate connections.

**Definition 4.3.** Candidate peers/connections. A device's neighbors who possess the content that the device requests formulate its candidate peers. A device and its candidate peers compose its candidate connections.

The candidate graph is a weighted directed graph $\mathbb{G} \triangleq \{N, E, W\}$, where node set N formulates the vertex set and $\mathbb{E} \triangleq \{(i,j)^k | e_{ij}^k = 1, \forall i, j \in \mathbb{N}, m_k \in \mathbb{M}\}$ is the edge set where $e_{ij}^k = 1$ if and only if device $i$ and $j$ formulate a candidate connection, where device $i$ transmits content $c_k$ to device $j$. The weight $W \triangleq \{w_{ij}^k | (i,j)^k \in \mathbb{E}\}$. In the proposed QoE-driven cooperative multimedia content dissemination scheme, MOS is the motivation of cooperation. Thus the connection MOS is labeled as the weight of the corresponding edge in the candidate graph, i.e., $w_{ij}^k = MQ_{ij}^k$. Figure 4.3 describes a snapshot of a D2D content dissemination system.

### 4.2.3  Feasible Graph Model

In the candidate graph, there may be candidate connections whose transmitter MOS and/or receiver MOS are so low that if they are selected as D2D connections, the total user satisfaction may sharp down. Therefore, these connections should be

**Fig. 4.3** An illustration of
graph model for a D2D
system

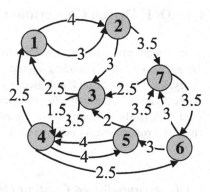

**Fig. 4.4** $\tau$-feasible graph for
Fig. 4.3

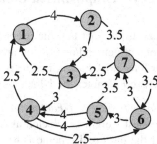

removed from the candidate graph. Additionally, there may also be connections
which posses the same transmitter and receiver while different contents. Under this
circumstance, the connection with highest MOS is the most feasible connection
and the other connections are redundant. Accordingly, acceptable connections and
feasible graph are introduced [1].

**Definition 4.4.** $\tau$-acceptable connection (set). A connection is $\tau$-acceptable if the
connection MOS is higher than the score threshold $\tau$, i.e.,

$$MQ_{ij}^k \geq \tau.$$

Then the connection of $e_{ij}^k$ is $\tau$-acceptable. Otherwise, i.e., $MQ_{ij}^k \leq \tau$, the connection
is $\tau$-unacceptable. If all the connections in the set are $\tau$-acceptable, the connection
set is $\tau$-acceptable.

**Definition 4.5.** $\tau$-**Feasible graph**. If a graph which satisfies the following: 1) each
edge is $\tau$-acceptable, and 2) there is at most one edge from $i$ to $j$ for all $i, j \in$ V.

Actually, a $\tau$-Feasible graph is formulated by removing $\tau$-unacceptable connections
and redundant connections from the edges of the candidate graph, i.e., feasible graph
is a subgraph of candidate graph. Figure 4.4 gives the 2-Feasible graph for Fig. 4.3.

## 4.3   QoE-Driven Cooperation Stimulation

In the cooperative content dissemination scheme, the user receives its demanding content at the cost of sending a content to its neighbor. From the view of graph theory, the out-degree and in-degree of each vertex is constant (one) in its $\tau$-Feasible graph. As a result, the contents are disseminated based on the 1-Factor of the feasible graph $\mathbb{G}_\tau$. In this section, we firstly analyze the 1-Factorable property for graph. Then we propose the cooperative content dissemination scheme.

### 4.3.1   Graph-Based Content Dissemination

Due to the fact that the users randomly distribute in the interested area and their content demand/possession are arbitrarily, the feasible graph may be 1-Factorable or have no 1-Factor, i.e., there are two cases: (1) $\mathbb{G}_\tau$ is 1-Factorable, i.e., $F \neq \varnothing$, where $F$ denotes the 1-Factor set for $\mathbb{G}_\tau$, and (2) $\mathbb{G}_\tau$ has no 1-Factor, i.e., $F = \varnothing$.

*Case 1: $\mathbb{G}_\tau$ is 1-Factorable* When $\mathbb{G}_\tau$ is 1-Factorable, the 1-Factor with the highest weight is selected as content dissemination scheme because the objective of the proposed scheme is to maximize the total user MOS. Thus, the problem is expressed as

$$\arg \max_{F_i \in F} w(F_i), \tag{4.8}$$

where $F_i$ is the 1-Factor of $\mathbb{G}_\tau$, $w(F_i)$ is the weight of graph $F_i$, representing the total user MOS. Actually, in order to solve the optimization problem Eq. (4.8), 1-Factors of $\mathbb{G}_\tau$ is the key point. Based on *Lemma 4.1*, the 1-Factors can be found by searching the permanence components of the adjacency matrix for $\mathbb{G}_\tau$. In order to explain the process clearly, we take the 2-feasible graph Fig. 4.4 as an example.

*Example 4.1.* The adjacency matrix of graph $\mathbb{G}_2$ in Fig. 4.4 is

$$\mathbf{A} = \begin{pmatrix} 0\,1\,0\,0\,0\,0\,0 \\ 0\,0\,1\,0\,0\,0\,1 \\ 1\,0\,0\,1\,0\,0\,0 \\ 1\,0\,0\,0\,1\,1\,0 \\ 0\,0\,0\,1\,0\,0\,1 \\ 0\,0\,0\,0\,1\,0\,1 \\ 0\,0\,0\,0\,0\,1\,0 \end{pmatrix}$$

The permanence of $\mathbf{A}$ is

$$p(\mathbf{A}) = a_{12}a_{23}a_{34}a_{41}a_{57}a_{65}a_{76} + a_{12}a_{23}a_{31}a_{45}a_{54}a_{67}a_{76} = 2. \tag{4.9}$$

Fig. 4.5 1-Factors for
Fig. 4.4. (a) $F_1$ and (b) $F_2$

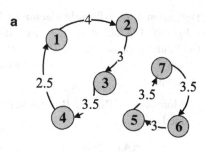

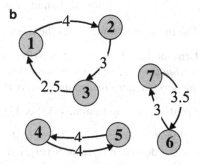

The result shows that the graph $\mathbb{G}_2$ has two 1-Factors, as shown in Fig. 4.5. From Eq. (4.9), 1-Factors of the graph include two spanning subgraphs $F_1, F_2$, i.e., $F = \{F_1, F_2\}$. The edge sets of them are $E(F_1) = \{e_{12}, e_{23}, e_{34}, e_{41}, e_{57}, e_{65}, e_{76}\}, E(F_2) = \{e_{12}, e_{23}, e_{31}, e_{45}, e_{54}, e_{67}, e_{76}\}$, respectively. Then

$$\begin{cases} w(F_1) = w_{12} + w_{23} + w_{34} + w_{41} + w_{57} + w_{65} + w_{76} = 20 \\ w(F_2) = w_{12} + w_{23} + w_{31} + w_{45} + w_{54} + w_{67} + w_{76} = 24 \end{cases}$$

Therefore

$$\arg\max_{F_i \in F} w(F_i) = F_2,$$

which denotes the content is disseminated according to $F_2$, i.e., node 1 sends content to 2, 2 to 3 and 3 to 1, node 4 and 5 send content to each other, and the same between node 6 and 7.

*Case 2: $\mathbb{G}_\tau$ has no 1-Factor* When $\mathbb{G}_\tau$ has no 1-Factor, some nodes can't send or receive content. In this case, $\theta$-*Pseudo-Factor* of the graph is introduced for content dissemination [1].

**Definition 4.6.** $\theta$**-Pseudo-Factor.** $\theta$-Pseudo-Factor of a graph $\mathbb{G}$ is a spanning subgraph of $\mathbb{G}$ including: 1) $\theta$ direct directed paths with distinct origin and terminus (including orphan nodes), 2) disjoint cycles which conclude all the vertices of $\mathbb{G}$ except the vertices in 1). $\theta$-Pseudo-Factors is a set of all the $\theta$-Pseudo-Factors of the graph.

Moreover, $\theta$-*Pseudo-Permanence* for the adjacency matrix of a graph $\mathbb{G}$ is defined as

$$\mathbb{P}(\mathbf{A}) = \{(a_{1j_1}, a_{2j_2}, \ldots, a_{nj_n}) | a_{1j_1} + a_{2j_2} + \cdots + a_{nj_n} = n - \theta,$$

$$\text{for all permutations } (j) = (j_1, j_2, \ldots j_n)\}.$$

The following lemma gives the method for searching $\theta$-Pseudo-Factor of a graph.

**Lemma 4.2.** *The amount of* $\theta$-*Pseudo-Factors for a directed graph equals to the cardinals of* $\theta$-*Pseudo-Permanence of its adjacency matrix and each component of the permanence corresponds to a* $\theta$-*Pseudo-Factor.*

*Proof.* Similar to Lemma 4.1. See [1] for detailed proof.                    ∎

Let $N^+(\mathbb{G})$ and $N^-(\mathbb{G})$ denote the amount of vertices which have no in-degree and out-degree in $\mathbb{G}$, respectively. Let $\lambda = \max(N^+(\mathbb{G}), N^-(\mathbb{G}))$. From *Lemma 4.2*, we have the following proposition:

**Proposition 4.1.** $\mathbb{G}$ *has no* $\theta$-*Pseudo-Factor if* $\theta < \lambda$,.

*Proof.* If $\mathbb{G}$ has $\theta(\theta < \lambda)$-Pseudo-Factor, then there are only $\theta$ direct directed paths with distinct origin and terminus. In other words, there are at most $\theta$ nodes which have no in-degree or out-degree. It is contrary to the condition that there are $\lambda > \theta$ nodes having no in-degree or out-degree. Thus, the proposition follows.          ∎

**Corollary 4.1.** *When* $v \geq \lambda$, $\mathbb{G}$ *has the minimum direct directed paths with distinct origin and terminus in* $\lambda$-*Pseudo-Factors among all its* $v$-*Pseudo-Factors.*

As $\lambda$-Pseudo-Factors has the minimum direct directed paths, we choose it as the content dissemination strategy in order to coordinate the most nodes to send content as well as receiving content. Then the problem is presented as

$$\arg \max_{P_i \in \mathbb{P}_\lambda(\mathbf{A})} w(P_i), \tag{4.10}$$

where $\mathbb{P}_\lambda(\mathbf{A})$ denotes the $\lambda$-Pseudo-Permanence of $\mathbf{A}$ and $w(P_i)$ denotes the weight of the graph corresponding to $P_i$. It is worthy noting that even all the nodes have in-degree and out-degree, the graph may have no 1-Factor. In this case, $\max(N^+(\mathbb{G}), N^-(\mathbb{G})) = 0$. In order to include this circumstance, we choose $\lambda = 1$ because the 1-Pseudo-Factors has the minimum direct directed paths. Consequently, we set

$$\lambda = \max(1, \max(N^+(\mathbb{G}), N^-(\mathbb{G}))).$$

for unified expression. We also give an example in this case.

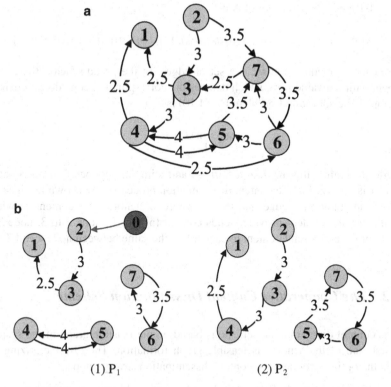

**Fig. 4.6** An illustration of (**a**) graph having no 1-factor and (**b**) the corresponding QoE-driven content dissemination graph

*Example 4.2.* Figure 4.6a is a 2-feasible graph for a D2D system. As node 1 and 2 has no out-degree and in-degree, respectively, $N^+(\mathbb{G}) = 1, N^-(\mathbb{G}) = 1$. Then $\lambda = 1$ and the graph has no 1-Factor. The content dissemination mechanism is designed by seeking its 1-Pseudo-Factors.

The adjacency matrix of the graph is

$$
\mathbf{A} = \begin{pmatrix}
0 & 0 & 0 & 0 & 0 & 0 & 0 \\
0 & 0 & 1 & 0 & 0 & 0 & 1 \\
1 & 0 & 0 & 1 & 0 & 0 & 0 \\
1 & 0 & 0 & 0 & 1 & 1 & 0 \\
0 & 0 & 0 & 1 & 0 & 0 & 1 \\
0 & 0 & 0 & 0 & 1 & 0 & 1 \\
0 & 0 & 0 & 0 & 0 & 1 & 0
\end{pmatrix}
$$

The 1-Pseudo-Permanence of $\mathbf{A}$ is

$$\mathbb{P}(\mathbf{A}) = \{(a_{12}, a_{23}, a_{34}, a_{41}, a_{57}, a_{65}, a_{76}), (a_{12}, a_{23}, a_{31}, a_{45}, a_{54}, a_{67}, a_{76})\},$$

where each element of the set is a set of edges. In the pseudo factor, there is an element with the value zero, which denotes the corresponding edge doesn't exist in the graph. The solution of problem Eq. (4.10) is

$$\arg\max_{P_i \in \mathbb{P}(\mathbf{A})} w(P_i) = P_1.$$

In order to realize the maximum fairness and stimulate cooperation, the content provider is involved in the content dissemination processes. As shown in Fig. 4.6b, we add an additional edge $e_{01}$ in $P_1$, where 0 denotes the content provider. Specifically, the content provider sends content to node 2, node 2 to 3, node 3 to 1, and node 4 and 5 send content to each other, the same between node 6 and 7.

### 4.3.2   The Cooperative Content Dissemination Scheme

As the content dissemination scheme is based on the feasible graph of the system, the first step of the scheme is feasible graph formation. Then by factorizing the graph, the optimal cooperative content dissemination strategy appears.

*Step 1 Feasible Graph Formation* Feasible graph formation is performed by iterations. At the beginning of each iteration, all users $i \in \mathbb{U}$ broadcast their cache status $\mathbf{X}^i$ to its neighbors and estimate SNR for the channel with its neighbors. Then they setup their transmitter MOS set $\mathbb{T}^i$ and receiver MOS set $\mathbb{R}^i$ by performing **Algorithm 1**. Upon receiving the MOS sets from the users, the CP formulates $\tau$-Feasible graph $\mathbb{G}_\tau$ and generates its adjacency matrix $A$ .

*Step 2 Cooperative Content Dissemination* The CP firstly judge whether $\mathbb{G}_\tau$ is 1-Factorable by computing permanence $p(\mathbf{A})$. If $p(\mathbf{A}) > 0$, the CP 1-Factorizes the graph and solves (4.8) for getting content dissemination scheme. Otherwise, CP computes $\lambda = \max(1, \max(N^+(\mathbb{G}), N^-(\mathbb{G})))$ and seeks $\lambda$-Pseudo-Permanence for $\mathbf{A}$. By solving (4.10), the CP constructs the content dissemination scheme. **Algorithm 2** in Table 4.3 gives the procedures of the scheme performed by the CP, where output $S$ is the edge set. In fact, an edge in $\mathbb{S}$ denotes a connection.

## 4.4   Discussions

This section analyzes security properties of the proposed scheme as well as proposing performance improvement strategy.

**Table 4.3** Algorithm 2: Cooperative content dissemination

| |
| --- |
| 01:  **Input:** $\mathbb{G}_\tau, \delta, \mathbf{A}, c_i$, for all $i \in \mathbb{N}$; |
| 02:  **Output:** $\mathbb{S}$; |
| 03:  **Procedures** |
| 04:  Compute the permanence $p(\mathbf{A})$ for $A$; |
| 05:  if $p(\mathbf{A}) > 0$ |
| 06:    Seek the 1-Factors $F = \{F_1, F_2, \ldots, F_{p(\mathbf{A})}\}$ for $\mathbb{G}_\tau$; |
| 07:    Solve the problem, $\arg\max_{F_i \in F} w(F_i) = F_k$, |
| 08:    $\mathbb{S} = \{(i,j) | (i,j) \in F_k\}$; |
| 09:  else |
| 10:    $\lambda = \max(1, \max(N^+(\mathbb{G}), N^-(\mathbb{G})))$, |
| 11:    Compute the $\lambda$-Pseudo-Permanence $\mathbb{P}(\mathbf{A})$ for $A$:; |
| 12:    Seek $\lambda$-Pseudo-Factors $P = \{P_1, P_2, \ldots, P_{|\mathbb{P}(\mathbf{A})|}\}$ for $\mathbb{G}_\tau$; |
| 13:    Solve the problem, $\arg\max_{P_i \in P} w(P_i) = P_k$, |
| 14:    $\mathbb{S} = \{(i,j) | (i,j) \in P_k\}$; |
| 15:  endif |

*Resistant to Free-Riding Attacks* In *case 2* of Sect. 4.3.1 when the feasible graph is pseudo-factorized, there are users who receive contents without serving others. Evidently, this behavior destroys fairness and discourages cooperation. However, if the devices, failing to work as a server at some iteration, are prohibited to receive contents, they may never have sufficient contents to share with others. Then the system availability will be reduced. Therefore, a debt mechanism is introduced in the system to address this issue. Specifically, at each iteration the content provider refreshes all nodes' debt record, denoted by $d_i$ for node $i$, as following:

- For the nodes who only receive content, the debt record increases one.
- For the nodes who only work as server, the debt record decreases one.
- For those who work both as senders and receivers, the credit record is not altered.

The content provider sets an integer $\delta > 0$ as the threshold. If $d_i < \delta$, node $i$ is permitted to receive contents. If $d_i \geq \delta$, node $i$ is permitted to sent contents but prohibited to receive anything until its debt falls below the debt line $\delta$. In this way, all the users are restrained to obtain $\delta$ free services at most.

*Cheat-Proof* As the users want to get as much media as possible while not willing to send contents to others, some devices may propagate fault MOS or possession status for their own benefits. The proposed scheme is able to realize possession status cheat-proof and MOS cheat-proof.

- **Possession status cheat-proof**. Usually, the cheaters may broadcast false status in two ways: (1) one claims that he lacks some content, but indeed he possesses; (2) one declares that he has some content, but in fact he lacks. In the first case, the cheater's neighbor may send him the content that he already possesses. This content sharing process is meaningless to the cheater, and even worse, it wastes his opportunity for obtaining demanding content. Accordingly, this cheating action does nothing but harm to the cheater. Thus, rational users will not commit this behavior. Similarly, if a user declares possession of his lacking content, this cheating behavior will be discovered once he is required to send the content. Consequently, the proposed scheme realizes possession status cheat-proof.

- **MOS cheat-proof**. Generally, a cheater may forge its MOS by rising up or descending the MOS on the true value. Specifically, if the connection with false MOS is high enough to be chosen as the D2D connection, the other really high MOS connection will be deleted at this iteration because each device has only one connection from and to other devices at each iteration. Consequently, this behavior will bring MOS loss to the cheater. Contrarily, if the cheater lowers down the MOS, the related connection may be deleted while another connection with lower MOS will set up. This connection will bring him bad quality of experience. Therefore, rational users have no motivations to forge the MOS.

*System Availability* Consider the circumstance that if the node obtains all the contents it demands (usually this happens near the end of the content dissemination processes), then it needn't exchange content with others anymore and then it may decrease its transmitter MOS for avoiding sending content to others, which will reduce the system availability. To solve this problem, we introduce *adaptive debt line* instead of a determined one. In specific, at the beginning $\alpha$ iterations, the debt line is set at a high level so that the nodes are able to obtain the contents from the source or neighbors without exchanging resources as most of them may lack most of the contents. After $\alpha$ iterations of the content dissemination, most of the contents are widely distributed among the nodes, then the debt line comes down and the nodes are compelled to share contents. The value of $\alpha$ and its descending speed is affected by the scale of the network and the quantity of the contents.

*Implementation in Large-Scale D2D Systems* When $n$ goes large, it is not easy to factorize the graph. In order to address this issue, we propose to divide the users into small groups, where the propose content dissemination scheme is implemented. In particular, the groups may be formulated according to the users' distribution or possession status. Notably, the D2D network topology varies frequently with users following various mobility directions and velocities. Accordingly, the member and even the amount of groups change frequently. The new groups are formulated according to the locations of the nodes and their possession status at the begin of each iteration. *Scenario 2* in Sect. 4.5 investigates the performance of the proposed scheme implemented in large-scale network environments.

## 4.5   Numerical Results

In this section we evaluate the performances of the proposed content dissemination scheme in realistic D2D communication scenarios. Additionally, we compare it with two alternative schemes: direct content dissemination scheme (DDS) and Social trust and social reciprocity-based cooperative content dissemination (SSCD) in [60] of Chap. 3.

- **Direct content dissemination scheme (DDS)**. All the users get the contents directly from the content provider and the content disseminations are implemented by iterations. At each iteration, the content provider serves six nodes with higher MOS compared with their counterparts.
- **Social trust and social reciprocity-based cooperative content dissemination (SSCD) in [60] of Chap. 3**. In SSCD, a coalitional game-theoretic framework is developed to devise social-tie-based cooperation strategies for D2D communications. The contents are disseminated according to users' social ties and the preference list, which is sorted based on their MOS.

### 4.5.1 Simulation Settings

*MOS Estimation Model* Motivated by Khan et al. [2] and Khan et al. [7], in our brief the MOS is estimated by Zhang et al. [1]

$$RQ_i^j(c_k) = 3.956 + 0.0919\ln(r_i^j) + CT^j(c_k) * (-5.8497 + 0.9844 * \ln(r_i^j)), \quad (4.11)$$

and

$$TQ_j^i(c_k) = 3.956 + 0.0919\ln(r_i^j) + CT^i(c_k) * (-5.8497 + 0.9844 * \ln(r_i^j)) \quad (4.12)$$

where $CT^i(c_k), CT^j(c_k) \in [0, 1]$ denote the content type of $c_k$ for user $i$ and $j$, which are determined by the users themselves. $r_i^j$ is computed from Eq. (4.4), where $\mu_i^j$ is estimated with the same method as in [60] of Chap. 3.

*Parameter and environment settings* Totally $n \geq 2$ devices with communication radius of 250 m are randomly deployed in an interest area of $500 \times 500$ m. We assume that all the users want to obtain all the 10 contents and each user stores 3 contents at the initial stage. The debt line $\delta = 3$. We complete simulations in two scenarios: (1) *Scenario 1*: scarcity network, where $n = 7$. In SSCD, the seven nodes have the same social trust with those in [60] of Chap. 3. (2) *Scenario 2*: density network, where $n = 35$.

### 4.5.2 Simulation Results

We compare the average MOS and MOS standard deviation among the three schemes, i.e., Proposed scheme, DDS and SSCD, in both scenarios.

*Scenario 1* Figure 4.7a shows that the proposed scheme has higher average MOS than the other two schemes for five of the seven nodes. Thus, the proposed content dissemination may bring better user experience than the other schemes for most of the nodes. Meanwhile, we evaluate the average MOS at each iteration in Fig. 4.7b.

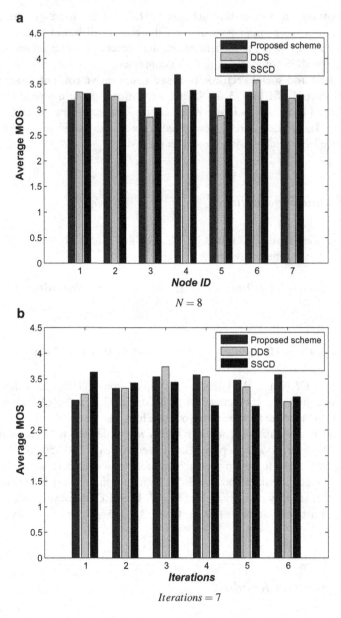

**Fig. 4.7** Average MOS varies with (**a**) the node and (**b**) the iterations in *Scenario 1* ($\delta = 3$)

Note that the three schemes may perform different iterations for disseminating all the contents. We choose the minimum amount of iterations for comparisons. From Fig. 4.7b we find that the average MOS varies at random. This is due to the fact that the nodes move randomly around resulting changing neighbors. Generally, the

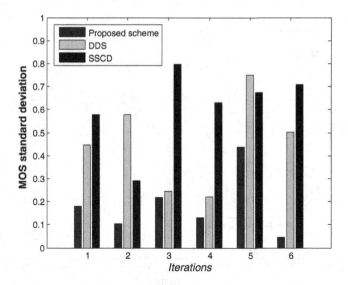

**Fig. 4.8** MOS standard deviation varies with iterations in *Scenario 1*

proposed scheme outperforms the other two except that sometimes SSCD obtains highest MOS, i.e., at iteration 1 in Fig. 4.7b. This is because the social trusted nodes may be physical proximity, bringing high QoE. Similarly, sometimes most of the nodes may locate near to the content provider thus the DDS may experience highest MOS, i.e., at iteration 3 in the figure. We also evaluate the MOS standard deviation to explore the fairness characteristics of the scheme. Figure 4.8 demonstrates that the proposed scheme has much lower deviation than the others. The encouraging results show that the proposed scheme has high fairness.

*Scenario 2* Figure 4.9 compares the average MOS and MOS standard deviation in a density network environment. The figure shows that the proposed scheme has the highest average MOS for most of the nodes at most of the iterations. This is not surprising because in large-scale network there is high probability to establish D2D links between nodes with density node distribution. Moreover, the figure also demonstrates that direct content dissemination may be intolerable as the MOS falls below 2. Consequently, D2D communications have superiority for multimedia services in large-scale networks. As users' opportunity of selecting servers and clients varies largely with the network topology, the MOS standard deviation fluctuates markedly in *Scenario 2*. Interestingly, the DDS scheme has the smallest deviation because it usually offers almost equal chance to all the nodes for content dissemination. Whereas, SSCD scheme has largest deviation in both scenarios owing to its social-tie-based content dissemination.

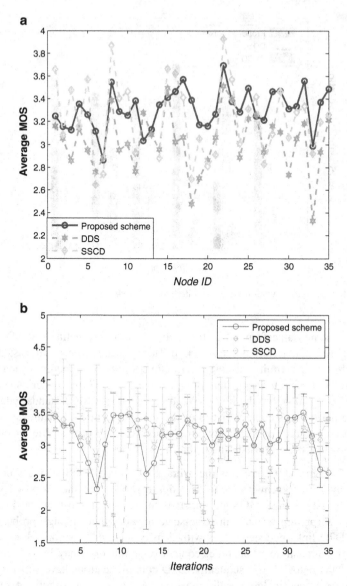

**Fig. 4.9** Simulation results of *Scenario 2* (**a**) Average MOS varies with node ID and (**b**) MOS standard deviation varies with iterations ($\delta = 3$)

## 4.6 Conclusions

In this chapter, a cooperation stimulation strategy for multimedia content dissemination in D2D communications is proposed. The scheme is designed based on graph theory and convex optimization aiming to maximize uses' QoE. Specifically, all users compute their transmitter MOS and receiver MOS by estimating the channel state information. After receiving the MOS sets from the users, the CP formulates a weighted directed graph according to the network topology and connection MOS. By factorizing the graph, the content dissemination scheme is designed. Additionally, a debt mechanism is introduced to resist MOS or content possession status cheating attack. Simulation results demonstrate that the proposed scheme gives due consideration to efficiency and fairness for content dissemination in D2D communications.

## References

1. Zhang A, Chen J, Zhou L, Yu S, Graph Theory based QoE-driven cooperation stimulation for content dissemination in device-to-device communication. IEEE Trans Emerg Topics Comput, to appear. doi:10.1109/TETC.2015.2430816
2. Khan A, Sun L, Ifeachor E (2012) QoE prediction model and its application in video quality adaptation over UMTS networks. IEEE Trans Multimedia 14(2):431–442
3. Host-Madsen A, Zhang J (2005) Capacity bounds and power allocation for wireless relay channels. IEEE Trans Inf Theory 51(6):2020–2040
4. Chung S, Goldsmith AJ (2001) Degrees of freedom in adaptive modulation: a unified view. IEEE Trans Commun 49(9):1561–1571
5. Chen WK (1965) Flow graphs: some properties and methods of simplification. IEEE Trans Circuit Theory 12(1):128–130
6. Tutte WT (1953) The 1-factors of oriented graphs. Proc Am Math Soc 4:922–931
7. Khan A, Sun L, Jammeh E, Ifeachor E (2010) Quality of experience-driven adaptation scheme for video applications over wireless networks. IEEE Trans Multimedia 4(11):1337–1347

# Chapter 5
# Summary

In this chapter, we summarize the main ideas of the brief and present the future research directions for security-aware D2D communications.

## 5.1 Summary of the Book

In this brief, we discuss several security issues for D2D communications underlaying cellular networks. In conclusion, the main ideas of the book are as follows:

- An overview of D2D communications is given. We overview D2D communications from the aspects of categories, characteristics, and application scenarios. Furthermore, we present the research topics, including physical-layer techniques, content dissemination, security, and cooperation stimulation. Their state of the art is reviewed and analyzed. Especially, we describe the security architecture of 3GPP ProSe, potential threats, and security requirements of D2D communications.
- A secure data sharing protocol is designed. In the proposed protocol, data confidentiality and integrity are assured by end-to-end encryption. Furthermore, if the transmitted data is not originated from the authorized provider or altered by some adversaries, the receiver is able to detect the event by signature verification and report a feedback message to the authority center. In addition, non-repudiation is realized by the checking record table in the eNB.
- A joint physical–application layer security framework is presented. With an overview of physical-layer security from the aspects of secrecy capacity, physical-layer key agreement, and physical-layer authentication, we consider both the physical-layer and application-layer security technologies for multimedia content delivery over D2D communications. By exploiting the

© The Author(s) 2016
A. Zhang et al., *Security-Aware Device-to-Device Communications Underlaying Cellular Networks*, SpringerBriefs in Electrical and Computer Engineering,
DOI 10.1007/978-3-319-32458-6_5

secrecy capacity and signal processing technologies at the physical-layer, and the authentication and watermarking strategies at the application-layer, the available network resources can be efficiently utilized without alterations of network architectures.

- A graph theory-based cooperative content dissemination scheme is proposed. In order to stimulate all the nodes to be more cooperative on content dissemination, each node is required to send content to others for gaining its desired content. In particular, the CP formulates a weighted directed graph according to the network topology and connection MOS. Then the candidate server and client for each node are chosen by factorizing the graph. The highest total user MOS is arrived by seeking the 1-factor with the largest weight. Notably, the scheme has the unique feature of easily implementation because the MOS of each candidate connection is distributively estimated by the relative nodes instead of a center.

## 5.2  Future Research Directions

Although there are researches concentrated on security issues in D2D communications, there are still numerous challenging problems and open issues in this area.

- *Cryptography solutions.* Cryptography approach is a typical security solution in modern communication systems. It is able to satisfy D2D communication security requirements in terms of data confidentiality, integrity, and entity authentication. Some security methods in other wireless networks, such as mobile ad hoc networks and wireless sensor networks, may be lent to security-aware D2D communications. However, due to the unique characteristics of D2D systems, it still faces challenges including key management, privacy preservation, and access control to achieve secure D2D communications.
- *Physical-layer security.* Physical-layer security is a hot topic in wireless communications recently, while its research in D2D communications is still in its infant. A significative direction worthy of investigation is the combination of physical-layer security with resource allocation, power control, and interference management. Another interest direction is to improve secrecy capacity of D2D communications by taking the advantages of interferences between the UEs and cellular users.
- *Joint physical–application layer security schemes.* Although a joint physical–application layer security framework is presented in this brief, we don't give a specific security solution to achieve security requirements in D2D communications. For future work, we may integrate physical-layer techniques into application-layer security mechanism, i.e., key agreement form the physical-layer for cryptography in the application-layer.
- *Game theory-based cooperation stimulation mechanisms.* Game theory has been widely used in networks for cooperation stimulation. Due to the fact rational people are involved in D2D communications, it is also a prospective method to

introduce game theory into D2D communication cooperation stimulate. The key points lie in formulating proper games and searching the optimal equilibrium.

- *Mobile social network assistant security-aware D2D communications.* Mobile social networks (MSNs) have attracted increasing attentions in research community and industry [1, 2]. D2D communication networks can be recognized as a mobile social network because the devices are hosted and controlled by the people. Consequently, the security and privacy solutions in MSN may be explored for D2D communications. However, it is a challenging work to construct social trust and social ties with privacy preservation for the devices with physical proximity, which remain open issues.

# References

1. Liang X, Lu R, Lin X, Shen X (2014) Security and privacy in mobile social networks. Springer, Heidelberg
2. Zhang Y, Pan E, Song L, Saad W, Dawy Z, Han Z (2015) Social network aware device-to-device communication in wireless networks. IEEE Trans Veh Technol 14(1):177–190

Printed in the United States
By Bookmasters